Legends in Their Own Time

A Century of American Physical Scientists

Legends in Their Own Time

A Century of American Physical Scientists

Anthony Serafini

Plenum Press • New York and London

Library of Congress Cataloging-in-Publication Data

Serafini, Anthony
    Legends in their own time : a century of American physical
  scientists / Anthony Serafini.
        p.   cm.
    Includes bibliographical references and index.
    ISBN 0-306-44460-7___
    1. Physicists--United States--Biography. 2. Astronomers--United
  States--Biography. 3. Chemists--United States--Biography.
  I. Title.
  QC15.S437   1993
  500.2'092'2--dc20
    [B]
                                                          92-43949
                                                             CIP

ISBN 0-306-44460-7

© 1993 Anthony Serafini
Plenum Press is a division of Plenum Publishing Corporation
233 Spring Street, New York, N.Y. 10013

Printed in the United States of America

Preface

America is a society that encompasses and reflects the accomplishments of science. As civilized people, we can argue that to understand America is to understand the scientific genius of the men and women who built the foundations of modern science and technology. However, science is not merely a collection of the results of scientific research or intellectual wizardry. It is also a social phenomenon with real human beings with all their foibles. As George Sarton said, "The best way to explain American achievements is to focus the reader's attention upon a few of the leading scientists." That I have tried to do. While there are other valid approaches to the history of science, certainly the biographical approach brings to light the fact that science is not a mere congerie of facts and theories. It is a very human activity with biases, personal rivalries, censorship, and even thievery, as the case of Rosalind Franklin shows.

A note is in order on the point in time I've chosen for the beginning of this book. I begin roughly in the late 1830s, in the Bond era in astronomy. It was in this period that American science education was beginning to come into its own—a sound educational system in science being a *sine qua non* for continual, systematic progress in scientific research. Before this period, science professors appeared on faculties only occasionally, often teaching several other subjects as well. By the 1830s, however, the professor of physics or chemistry appeared in college catalogues as often as did professors in the humanities. Also, specific courses and specialists were appearing in the 1830s, including even such "arcane" fields as geology. Wesleyan University, for example, could boast of three scientists on a faculty of only seven. By the 1850s, Amherst College had more faculty in science than in any other field.

The period between 1875 and 1910 is crucially important as well. As historian Stanley Guralnick has pointed out:

> The period during which the sciences became fully established in academe, 1875–1910, witnessed the rise of multipurpose universities, development of the graduate school, and the division of the faculty into separate departments—changes that have given the essential intellectual and administrative character to our present university.

It was during these periods that an increased awareness developed of the need for stressing the theoretical underpinnings of applied science. "Backyard tinkerers" had produced many advances, but the scientific community had now begun to fully appreciate the fact that real scientific advance could not be made without systematic study of scientific theory. In short, science was becoming a profession.

Later, the lone Byronic heroes of American science began to receive institutional backing of various kinds. An important forward step came with the Morrill Land Grant College Act of 1862. Recognizing the need for trained scientists rather than "tinkerers," this act led to the founding of Cornell University in 1865 as well as to the founding of several other institutions. Two decades later, after the Morrill Act of 1882, the American Association for the Advancement of Science appeared. For some of the sciences, such as physics, however, there was really no professional society until the founding of the American Physical Society in 1899.

By the early decades of the 20th century, science was developing in three arenas: the growth of corporate laboratories, universities, and governmental institutions. As the public grew more comfortable with science, this confidence soon led to increased funding. In the early decades of the century, government aid to science was minimal. During the First World War, however, organizations like the National Research Council and the Carnegie Foundation began to appear, ready to infuse money into scientific research. It was becoming evident that science had an obvious relevance to the war effort—the development of chemical weapons being one such example. Over the course of the next few decades, other organizations such as the Russell Sage Foundation and the Guggenheim Foundation added their resources.

Keeping in mind the principle that no selection can be completely "objective," I have followed Sarton's advice in selecting, as he says, "a few of the leading scientists." The problem then, of course, is to decide who should be covered and by what criteria they should be selected. Although men like Michelson and Rowland are rather obvious choices for this book, there are many others. I would have liked to have been able to devote more space to scientists such as Simon Newcomb, Wilhelmina Fleming, Theodore Richards, Richard Chace Tolman, and John Trowbridge, among many others.

The problem of selection compounded itself exponentially with the explosion of scientific research after the embarrassment of *Sputnik* in 1957. I might have covered, for example, Cornell physicist Ken Wilson's work in phase transitions, for which he captured the 1982 Nobel Prize. And in astronomy, there is Jocelyn Bell's discovery of the first pulsar, *CP 1919*, in 1967, which led to speculations that extraterrestrials were sending messages to Earth—a speculation refuted only when astronomer Thomas Gold proved that the "message" was merely a quickly rotating neutron star.

Then there is the wonderful story of the development of the *laser*, dating arguably from Charles Townes's work in the 1960s on the principle of the laser. And by 1981, Arthur Schawlow along with Kai Siegbahn of Sweden would capture the Nobel Prize for applying the laser to studies of the structure of solid-state substances.

In this respect, this book makes no claim to being comprehensive. Some "decision procedure" on what and whom to cover was therefore necessary. Where choices were balanced, I tended to side with "pure" science rather than technology, with a few exceptions—hence, the omission of the development of the laser. I've also covered those scientists whose work best illustrates the "pioneering spirit"—the urge to explore new and uncharted territory and the great sense of practicality associated with that spirit. Nineteenth century scientists like Michelson and Gibbs had, for example, fewer of the normal "support systems" we take for granted today. There were few academic journals, and communication with other scientists was more difficult and slower than it is today. Furthermore, large grants and salaried positions were few.

Irving Langmuir of the General Electric Laboratories in New York could hardly be omitted from a work like this. Beyond his brilliant work in both experimental and theoretical science, he represented a new breed in this country—the scientist working in an industrial laboratory.

In the end, I selected scientists as much for what their lives tell us about how science is done as for what they actually did. Of course, none of the individual profiles in this book pretends to completeness or thoroughness. A really thorough account of the scientific work of scientists like Pauling, Annie Cannon, and Irving Langmuir remains to be written.

A final note on the book: where possible, I have tried to include phases of a scientist's career that are perhaps a bit less well known. For example, in the case of Irving Langmuir, while his work on, say, the structure of the atom is well known, his courageous support of biochemist Dorothy Wrinch is mentioned only rarely in the literature. Also, Lawrence's pioneering work in medicine is perhaps less known than his work on the cyclotron, so his medical research is discussed here as well. Then, too, I have tried to bring out some recent research that argues persuasively that E. W. Morley's contributions to American science have often been overshadowed by his more famous collaborator, A. A. Michelson. The reader will note too that I have devoted considerable space to solid-state physics—a branch of that science that is, I think, too often overlooked in favor of the more "exciting" fields like particle physics and astrophysics.

The history of American science is the story of how all of these accomplishments came about since the nation-building and difficult decades of the 19th century, the period during which American science began to evolve into a profession.

I would like to thank the following people for looking over various portions of the manuscript and making suggestions: Spencer Weart, Director of the Center for History of Physics of the American Institute of Physics, Robert Olby of the University of Leeds, Robert Kargon of Johns Hopkins, Argonne National Laboratory, Katherine Olesko of Georgetown University, and Professor Paul Joss of the Department of Physics at MIT. I would also like to thank Professor Joss for many helpful conversations and for keeping me up to date on the latest work in particle physics and astrophysics. Additionally, I am indebted to John Archibald Wheeler, P. W. Anderson, and Verner Schomaker for offering their thoughts on Pauling's career.

I am grateful also to Lee R. Hiltzik, Archivist of the Rockefeller Archive Center in North Tarrytown, New York, for providing me with correspondence from Karl Darrow to Dr. Duncan MacInnes, and Professor David Miller of the University of California, Berkeley for his helpful comments and research material on Henry Rowland and for a copy of his doctoral thesis, "Henry Augustus Rowland and His Electromagnetic Researches," submitted to Oregon State University in 1970. (Interestingly, Professor Miller could speak with unusual authority on Rowland, as Rowland's daughter once took him to lunch for some "Baltimore Chicken," which Miller describes as "wonderful.")

The plan I followed in writing this book was approximately as follows. For the framework of the book, I relied heavily on scholarly articles by noted authors in the field, such as Robert Kargon, Spencer Weart, Daniel Kevles of Caltech, and many others. From these articles I was able to locate what appeared to be the most critical primary sources. For example, I consulted Millikan's autobiography, the biography of Michelson written by his daughter, as well as the book, *Light Waves and Their Uses*, by A. A. Michelson, University of Chicago Press, 1902. For the chapter on Arthur Compton, I drew heavily upon his own writings in *The Cosmos of Arthur Holly Compton* as well as his other writings. I would also like to thank Margorie Graham, Associate Librarian at the AIP, for providing me with a copy of Compton's writings and notes on cosmic rays, particularly those entries from June 1932, from the Compton notebooks.

Because this book is aimed at an educated lay audience rather than a group of specialists, I have tried to keep footnotes to a minimum and have adopted the admittedly unusual procedure of weaving some of the endnotes into the main text of the book. I have done this where it seemed that I could do so without unduly disrupting the flow.

For the chapters on astronomy (which, incidentally, I decided to break up into two in that quite different issues were involved), I wish to thank the Department of Space History of the National Air and Space Museum of the Smithsonian Institution. Because of their assistance, I was able to acquire the "Report on the History of the Discovery of Neptune," by Benjamin Gould (Smithsonian Institution Press, Washington, D.C., 1850), "On the Law of Vegetable Growth and the Periods of the Planets," by Benjamin Peirce, *Proceedings of the American Academy of Arts and Sciences*, 1852, 2:241, and "A Photographic Search for Planet O," *Annals of the Astronomical Observatory of Harvard College*, 1911. I also examined some of William Pickering's writings on file at Harvard. For this I am indebted to Patrice Donoghue, Curatorial Associate of the Harvard University Archives in the Pusey Library, who kindly sent me several articles from William Pickering's "A Search for a Planet Beyond Neptune" (Vol. LXI, Part II, *Annals of the Astronomical Observatory of Harvard College*). Also helpful were various British science museums and archives, which provided me with articles like "Account of some circumstances historically connected with the discovery of the planet exterior to Uranus," by George Biddell Airy, published in *Monthly Notices of the Royal Astronomical Society*, 1846, and "On an ultra-Neptunian planet," by George Forbes,

Proceedings of the Royal Society of Edinburgh, 1880. I relied additionally on published work by Owen Gingerich, Stanley Guralnick, Deborah Jean Warner, Carlene Stephens, and Katherine Olesko. I also relied on Mitchell Wilson's fine book, *American Science and Invention.*

For material pertaining to Henry Rowland, I wish to thank Joan Grattan, Manuscripts Assistant at the Rowland Manuscript Collection of the Milton Eisenhower Library at Johns Hopkins, who was kind enough to provide me with Joseph Ames's notes on Rowland's lectures on light in January–June 1888, several letters from Rowland to his family, letters from Rowland to Gilman, and Rowland's Rensselaer notebook of 1868. I also relied heavily on the 1902 edition of *The Physical Papers of Henry Augustus Rowland.* The Rowland Manuscript Collection of the Milton Eisenhower Library provided me with some photocopies of letters that Henry Rowland wrote to his family as well as with his correspondence with Daniel Gilman, president of Johns Hopkins in the Rowland era. They generously provided me with correspondence with Hermann von Helmholtz and some parts of his Rensselaer notebook of 1868 where Rowland discusses the source of the magnetic effects of electric currents.

For the Gibbs chapter, I went first to Lynde Wheeler's excellent biography, *Josiah Willard Gibbs, the History of a Great Mind.* Judith Ann Schiff, Chief Research Archivist at the Yale University Library, and Vincent Giroud, Curator of Modern Books and Manuscripts, directed me to the poem "The Storm" written by Gibbs in 1893 and held in the Gibbs Family Papers and two letters from Gibbs to Simeon Baldwin held in the Baldwin Family Papers in Manuscripts and Archives.

For the chapter on the work of Davisson and Germer, particularly the laboratory notebooks of L. H. Germer, as well as for archival material on solid-state physics, I wish to thank Dr. Sheldon Hochheiser, Senior Historian of the AT&T Archives, who kindly offered me access to the AT&T Archives (which includes the Bell Laboratory Archives) in Warren, New Jersey. For the chapter on Michelson, I wish to thank Marjorie Graham, Associate Librarian at the American Institute of Physics, for providing me with much helpful material, including biographical material on Michelson's student, William Kadesch. Also, Freiberger Library Special Collections of the Case Western Reserve University and the Rowland Manuscript Collection of the Milton Eisenhower Library at Johns Hopkins provided me with the article "The Relative Motion of the Earth and the Luminiferous Ether" by A. A. Michelson, *American Journal of Science*, 1881, as well as a copy of Michelson's papers on measuring the velocity of light, including "On a Method of Measuring the Velocity of Light," in the *American Journal of Science*, 1878. They also provided me with the article "On a Method for Making the Wave Length of Sodium Light the Actual and Practical Standard of Length" by A. A. Michelson and E. W. Morley, *Journal of the Association of Engineering Societies*, May 1887. Johns Hopkins also generously provided me with correspondence between Michelson and J. Willard Gibbs and with E. W. Morley.

I also wish to thank the Crerar Library of Science and Technology of the University of Chicago for providing material on Michelson. For Irving Langmuir, I relied heavily on the Sophia Smith Collection at Smith College for Langmuir's correspondence with a variety of scientists, including Linus Pauling and Dorothy Wrinch, as well as various reminiscences by friends and colleagues. Special thanks for the Sophia Smith Collection are due Bonnie Ludt, Administrative Assistant at the Institute Archives at Caltech, and M. Tess Legaspi, who generously provided me with microfilm information from the Robert A. Millikan Papers.

Contents

Legends in Their Own Time

A Century of American Physical Scientists

CHAPTER 1

Astronomy and the Harvard Observatory

Since the beginning of civilization, astronomy has been one of the most fascinating of the sciences. And it has been one of the most practical. In the days before modern technology, the stars were the only source of navigational information. So it was with the ancient Egyptians and the Phoenicians. In America, the practicality of astronomy has been evident. Farmers made use of the heavens to predict the weather while sailors predicted the motion of tides.

Americans have been interested in astronomy since colonial days. In 1769, David Rittenhouse was studying the orbit of Venus. By 1829, Nathaniel Bowditch became the first American accepted in the Royal Society of England. In 1842, Charles Dickens published his travel memoirs to the United States under the title *American Notes*—a scathing assault on what he viewed as American hypocrisy, medieval treatment of prisoners, and many other social ills. But even if it cost Dickens American readers, the message of *American Notes* could not be ignored.

The social reform momentum intensified when Dorothea Dix, in 1843, revealed the abysmal conditions in prisons and asylums, corroborating much of what Dickens had said. During the next decade, for example, many eastern states would pass laws prohibiting child labor in factories. This was significant for science: as is evident from a study of history, it is virtually impossible to pursue abstract study of any kind under wretched living conditions. Because of the reforming work of Dickens and others, reform bills were passed. Working hours and conditions improved and people had more leisure time for such "luxuries" as science. Because of this great change in the United States, American science was caught up in the reform movement as well. The Smithsonian Institution was founded in 1846 and five years later, Isaac Singer would revolutionize American life with the continuous-stitch sewing machine.

In this climate the first American astronomers flourished.

THE BOND ERA AT HARVARD

Luckily, President Quincy of Harvard always had a keen eye for new trends. Seeing a potential future for the field of astronomy, he approached William Bond in 1839. Since 1830, Bond and his son George Bond had run a small Boston factory named William Bond & Son, manufacturing timepieces for ships. Like so many scientists of the day, the Bonds were largely amateurs, having studied astronomy and physics on their own. William Bond and his son accepted President Quincy's offer to serve as the first two directors of the Harvard College Observatory and they would go on to serve in this capacity until 1865. It is hardly a surprise, then, that some historians take the founding of the Observatory as the real beginning of the growth of American astronomy.

In the early days of American astronomy, things were not easy. For example, a comet generated a regional furor when it appeared in 1843. The public besieged Harvard for information. But, according to the historian Carlene Stephens, because of the inferiority of the instruments then available to Harvard, their scientists were hard-pressed to come up with answers. (Even Bond's Dana House instruments, now on display at the Smithsonian, though arguably the equal of the best Germany could offer, were inadequate for the task: they could not accurately compute the comet's period, or the time it takes a comet to complete a single revolution of its orbit.)

Embarrassed, Harvard and Bond realized that an upgrading of the astronomy facilities was now inevitable. But Harvard was not the rich institution it is today. To accomplish his goals, Bond had to beg wealthy citizens in Massachusetts in order to collect enough money to install a 15-inch refracting telescope (similar to the one installed by Hale in the early days of Caltech) as well as a shelter for it. To do this, the Harvard faculty joined forces with a group of scientific dilettantes in Boston, including merchants and businessmen of every stripe.

Such citizens, from Nantucket to Salem and Lowell, came up with over $20,000. Businessman David Sears alone gave $5000. A proviso is in order here, however: this outpouring of interest in astronomy on this occasion was an exceptional occurrence— possibly due more to the prestige of Harvard and the special fascination of astronomy than to the popularity of science as such.

Outside of astronomy and this rather special event, the astronomer Charles Pickering was doubtless correct in saying: "the time is not at hand when scientific services shall be justly appreciated in the pecuniary way." And as historian of science Robert Post has convincingly argued, "perceptions of science tended to things ethereal [and] useless. . . ."[1] In short, science was misunderstood at all levels. Not only did most people perceive it as the idle pastime of the rich, but the public did not even understand what science *was*. For example, in 1849, *Beach's Periodical*, a leading popular, rather than academic magazine of the day, confused science and technology with abandon, constantly lumping together funds for a National Observatory with ordinary nonscientific construction projects, as if science and technology were one and the same.

Gradually, however, under Bond's guidance, Harvard succeeded. Bond had the good fortune to attract more and more talented men and the comparatively few female scientists of that time. In the years to come, Harvard began a gradual and inexorable rise to prominence in the field of astronomy and eventually astrophysics. Soon, in 1848, astronomers discovered the eighth moon of Jupiter and made pioneering studies of the nebulas in Orion. In 1858, they made critical studies of Donati's Comet, first discovered on June 2, 1858 by Italian astronomer Giovanni Donati. American astronomer Theodore Lyman performed epochal studies of the spectral lines of stars—the lines of varying intensity that are detected by a device called a "spectroheliograph." Since a line of a given width, intensity, and color is peculiar to a given chemical element, the pattern of lines tells the observer what elements are in the stars. (Today, the famous "Lyman series" of hydrogen lines is named in his honor.)

Later, Bond, as director of Harvard Observatory, was to play at least a minor role in the search for a planet beyond Neptune. Then, encouraged by some studies of British astronomer John Hind, he persuaded the superintendent of the U.S. Naval Observatory, Matthew Maury, to continue the quest. Yet he was not successful, and the matter was dropped for nearly 30 years. American astronomer Percival Lowell took up the quest but he too was unsuccessful. Clyde Tombaugh, an amateur astronomer who was helping out at the Lowell Observatory, took up the task. Finally, in February 1930, he discovered Pluto.

The other major professor of astronomy at Harvard was Benjamin Peirce. The latter is a justly honored name in the history of American astronomy. He was the first to hold the Perkins Endowed Chair, established in 1842 in honor of the distinguished 19th century philanthropist, Jacob Perkins, given only to the most distinguished and accomplished astronomers. Later he would serve as the Harvard consultant to Alexander Bache's Coast Survey. Bache was superintendent of the U.S. Coast Survey, charged with determining longitudes along the then-uncharted coastline of the eastern United States from Baltimore to Washington. According to historian Stanley Guralnick, Peirce was an elitist, who supported the elective system at Harvard merely to turn the poorer students away from his courses.[2]

Interestingly, Bond and Peirce were not always on the best of terms. According to historian of science Deborah Warner, although Bond had contributed significantly to studies of the rings of Saturn, Peirce failed to give him his due in a paper Peirce later delivered on that topic.[3]

THE WHIPPLE ERA

Then there was Fred Whipple, a Boston-based astronomy enthusiast who took some of the earliest photographs of the moon. He took his first low-quality daguerreotypes in October 1847. In 1849, he tried again, getting photos that showed a fair amount of detail of the moon's surface, considering the technology of the age, though not all historians agree with this claim. Carlene Stephens, for example, argues that Whipple's 1849 photos were not successful. In any case, there is general agreement that on July 16 and 17 of 1850, he did adequately photograph the star Lyrae.[4]

There was some humor in all of this. During that initial series of moon photos in December 1849, astronomer William Bond made the following entry in his log book: "On the evening of the 18th, just as we were commencing observations on Mars, Messrs. Whipple and [Arthur] Jones came to take a dag [daguerreotype] of the moon. Very much against our inclinations, we unscrewed the micrometer. . . ."[5] Apparently, the explanation of this reluctance was the fact that in some earlier work, in October 1847, Whipple had taken a photo of the moon through the refractor, although the sun's rays had burned a hole in Bond's coat! Presumably, Bond did not trust Whipple's manual dexterity.

Buoyed by the initial success of his first series of photos of the moon, Whipple and Boston astronomer Arthur Jones returned several months later and scored another "first." They managed to take an adequate photo of a star so faint that it had been previously regarded as unphotographable. Still more progress came when, on March 2, 1851, the *Harvard Journal* announced: "[He] succeeded in daguerreotyping Jupiter about 11PM—took six plates more than 24 hours old—."[6]

His enthusiasm lifted still further, Whipple continued his photographic assault on the heavens, shooting innumerable pictures of the stars, moon, and numerous other objects over the next several months. But as before, the work was slowed by the comparatively poor quality of both the cameras and the telescope and, consequently, the photos. (Photos of the moon were much better in 1852 than in 1851.)

He soon discovered, however, that he could improve the quality if he focused the telescope so that the exposure occurred directly on a plate. In the earlier method, the image was allowed to impinge first on a silvered mirror, which then was "bounced" onto the photographic plate. By 1851, Whipple had learned that a shorter exposure time would actually give better results. In that year, they managed to obtain passable photos not only of Jupiter, but also of a solar eclipse.

Soon, another talented astronomer, Lewis Rutherfurd, then at Columbia, entered the ranks of the small-but-growing coterie of American astronomers. He started shooting the heavens in New York City in 1861, inspired by Draper's work in astronomical photography.

THE DRAPERS

By May 1872, Dr. Henry Draper, son of William Draper, had taken one of the first successful photographs of the Fraunhofer lines in a star other than the sun. These lines appear in a special "spectroscopic" photo, indicating what elements compose the star. The Fraunhofer lines are described very well by Harvard scientist Annie Cannon in her essay "Classifying the Stars":

> In 1814, more than a century after Newton, the spectrum of the Sun was obtained in such purity that an amazing detail was seen and studied by the German optician, Fraunhofer. He saw that the multi-

ple spectral tints, ranging from delicate violet to deep red, were
crossed by hundreds of fine dark lines. In other words, there were
narrow gaps in the spectrum where certain shades were wholly
blotted out.[7]

Although the spectral lines had been seen earlier, Draper was the
first to actually photograph them. But in his revolutionary work,
he pushed this kind of research to its limit—detecting several
such lines in more distant stars. With this kind of information,
scientists were able to learn much more about the kinds of elements in stars, including their age, history, and so forth.

And there was much more. Draper had been the first to show
that the spectrum of any solid substance heated to incandescence
was "continuous," meaning that there were no gaps in the visible
lines caused by the radiation. All colors, from infrared to ultraviolet, were in the spectrum. In other words, any material (e.g.,
copper, iron, salt) would, at that temperature, begin to give off
visible light, thereby becoming "incandescent." Draper had further proven that all solid substances become incandescent at about
977°F. He showed as well that the hotter a body became, the more
radiation it emitted, mostly in the form of invisible ultraviolet and
infrared light. This would later prove to be of colossal importance
when the first prototypes of infrared and ultraviolet telescopes
such as "Einstein" and "AXAF" (Advanced X-Ray Astronomy
Facility) appeared on the drawing boards in the 1960s.

As versatile a scientist as America ever produced, Draper was
also among the first to do systematic research on plant physiology.
In recognition of his contributions, the American Academy of
Science honored him in 1876 with the Rumford Medal, its highest
award, for his research in astronomy and, in particular, for his
work in radiant energy. And Charles Francis Adams, the president
of the Academy, in his presentation of the Rumford Medal, made a
point of listing virtually all of Draper's contributions.

Not surprisingly, astronomy gradually exerted more and
more of a grip on the public's imagination as the 19th century
passed. The school curriculum reflected this. Students began
clamoring for more astronomy courses at the college and even the

secondary school levels. In 1865 alone, four major universities were founded—the University of Maine, Cornell, Kentucky, and Purdue, all paying close attention to science. First among equals was Cornell with its new college of agriculture, the Rockefeller Physics Department, and some of the earliest practitioners of scientific psychology.

THE PICKERING ERA

Edward Charles Pickering was born in Boston on July 9, 1846. He received his bachelor's degree from the Lawrence Scientific School in 1865, where he then taught math for two years. In 1877, he assumed the directorship of the Harvard College Observatory. This event was significant in that it heralded a new era in astronomy in the United States: a shift from merely describing the locations and movements of heavenly bodies—"positional" astronomy—to what is now known as astrophysics—the application of the principles of physics to fundamental problems in astronomy. The challenge to Pickering was great, for others were already doing significant work. Two years after Pickering's appointment Italian astronomer Giovanni Schaparelli made his renowned claim that he had discovered canals on the surface of Mars.

By 1889, Pickering, already a full professor of physics at MIT, was instrumental in introducing laboratory work into the instruction of physics in the United States. Indeed, a laboratory manual he wrote was widely used for many years at MIT to teach laboratory experimentation. While Pickering was not shy (saying of his laboratory manual that it was "of direct practical value . . ."), the manual really was a significant work since, in this era, the modern phenomenon of textbook proliferation had not yet arrived. Good ones were few and far between. Even scarcer were laboratory manuals in physics, since very few schools offered laboratory work. After the laboratory manual, Pickering wrote his *Elements of Physical Manipulation*, a more theoretical approach to laboratory work, stressing why various laboratory procedures did what they did. It was published in 1873 and was again based on his own laboratory teaching at MIT.

Here it is worth noting that while much has been made of the often-touted superiority of German universities, schools like MIT, according to historian of science K. M. Olesko, at least compared well to German "mechanical institutes," if not to universities. And in some respects, as in the quality of their laboratory work, even the German universities of the 1870s were slow to catch up to American standards.[8]

Pickering's generosity was without peer. As historian of science William Koelsch points out, he even assisted the fabled A. A. Michelson, making it possible for Michelson to use the Harvard telescopic facilities. Still later, he nominated Michelson for the Nobel Prize.

He was a visionary as well, farsighted enough to appreciate that physics would be the basis for innumerable new professions, such as electrical, chemical, and mechanical engineering. Most schools, including his own, however, were not so visionary. Even as late as 1891, MIT's catalogue said "most of the students taking the course in physics intend to make teaching their profession."

In 1891, Pickering set up an observatory in the city of Arequipa in northern Peru, under the auspices of Harvard, and, three years later, the Flagstaff Observatory in Arizona, where in 1903 he created the first comprehensive map of the surface of the moon.

Yet, despite a stellar array of triumphs, when Pickering was nominated for the director's post at the Harvard Observatory, a great hue and cry emerged from the astronomical community. Although he did get the post, his orientation in physics rather than astronomy bothered many scientists. However, since he'd successfully incorporated physics into laboratory work as well as into his teaching of astronomy, he was reluctant to abandon physics when he took up the reins of the Harvard Observatory. To the legions of astronomers untrained in physics, this was a threat, pure and simple.

NEW TECHNOLOGY IN ASTRONOMICAL PHOTOGRAPHY

Using the approach of a precise experimental physicist, Pickering commissioned Alvan Clark, the distinguished telescope maker at Cambridge, to assist him in his astronomy research. Pickering had good reason to trust Clark's abilities: Clark had distinguished himself with such feats as locating in 1862 the companion star to Sirius. This he had done with a telescope of his own design and construction. Also, Clark managed to build a 36-inch telescope for Professor James Keeler, who used it to make pioneering studies of the composition of the rings of Saturn. In the 1850s, the firm of Alvan Clark & Sons was producing optical instruments to rival the finest German instruments, albeit aided, it must be said, by such fine German lens grinders as William Wurdemann. Among many other accomplishments, Clark had, by 1862, built a refractor with a 19-inch lens, larger even than the best German lens in Munich.

Clark, on the other hand, had reason to be wary; on at least one occasion, a lens he'd made for the University of Mississippi was never paid for or used. But though he was initially cautious, Clark and Pickering soon developed a satisfying working relationship. With Clark's inspiration, as well as his lens craftsmanship, Pickering went on to numerous other accomplishments, including the development, in 1882, of better techniques for photographing stars, involving the use of a large prism in front of the photographic plate.

MORE GROWTH IN SPECTROSCOPIC ASTRONOMY

Though Pickering's contributions were many, one in particular stands out: the use of photography in studying stellar spectra. From roughly 1839 to 1882, the old and primitive French system of photography, the daguerreotype, was still essentially the only technique for astronomical photography.

Not satisfied with the existing methods for photographing an ordinary visual image of a star, Pickering wanted to capture them as seen through a spectroscope. Pickering's technique involved positioning a large prism—the distinguishing feature of a spectroscope—on an ordinary telescope. That, in effect, converted the telescope into a powerful spectroscope. The prism, by breaking up the light into its component colors which are then more easily analyzed, offered a wealth of other information about the composition and temperature of the star—just as prism viewing had yielded such information about our own sun years earlier.

With such an arrangement, one exposure on a photographic plate would reveal a spectrum for every star in the photo. The reason is simply that each star is itself a point source of light. As such, that light, viewed through a prism, would be broken up into its various spectral colors. The result was a collation of the spectra of innumerable stars that Pickering and staff collected with continued and serious effort.

Shortly after Pickering's work, an English scientist, Sir William Huggins, showed that the quality of stellar pictures could be improved dramatically by using dry gelatin plates that he'd developed. And in 1868, Huggins, like the Drapers, used the spectroscope to study a number of nebulas seen by the French scientists Messier and Lord Rosse many years earlier.

As Huggins wrote in 1867, "I directed the spectroscope to one of the these small nebulae. The reader may now be able to picture the degree of awe with which . . . I put my eye to the spectroscope. A singular [image] only such as is given off by a luminous gas!"—ample testimony to the new worlds and wonders opened up by the spectroscope. Later, British astronomer Sir William Herschel would refer to these as "island universes," because of the huge number of stars and spectacularly radiant gas. The probing continued unabated. As scientists scanned the heavens, they found, on their photographic plates, increasing numbers of stars, nebulas, etc., which no one had previously noticed.

Why was photography such a turning point in astronomy? For one thing, Pickering's pioneering work in that area had made it possible for an observer to "see" more of the sky than anyone had previously. Also, the photo offered great flexibility for study. Unlike the momentary viewing of an image seen directly through the telescope, a photo could be sent to other astronomers for their opinions.

PICKERING'S SYSTEM OF CLASSIFICATION OF STARS

But progress on stellar spectra was not limited to the United States, of course. At around the same time or earlier, Angelo Secchi of the Observatory of the Collegio Romano in Italy had succeeded in dividing all stars roughly into four classes, based on his own studies of some star spectra. But Pickering even improved on that. Although Pickering was impressed with the schemata, he wished to make it even more precise and quantitative. Thus, he devised a system for grading stars that still influences contemporary astronomical practice (though his system as such does not exist in the same form).

The essence of the scheme was as follows. Pickering constructed 14 categories of stars—the first labeled "A" and the last labeled "N." The principle of ordering was ingenious: The spectral lines given off by hydrogen became his reference point, merely because they were simple in composition and easy to see. Stars with the strongest hydrogen spectral lines were placed in category "A"; those with the weakest were considered class "N." While the classification scheme was to some extent arbitrary, the fact remains that categories provided a method for objectively measuring the internal temperatures of stars. The reason is that the intensity of spectral lines varies directly with temperature: the more intense an object's spectral lines, the hotter the object is. Thus, the range was from the class O, white-blue stars, through yellow stars, down to the reds in class M.

Eventually, over the years, that kind of temperature information allowed astrophysicists to discover all sorts of other information about the makeup of stars, even leading to major theories as to how the stars and, in fact, the universe itself were formed. Pickering's work here proved to be critically important for later astronomical work. At the time, the accepted theory of the makeup of stars held that the sun was either a solid or a fluid body. Now, of course, we know that it is gaseous, incandescent (gives off visible light), and encircled by a huge incandescent atmosphere.

Naturally, as new data were collected, alterations had to be made in the classification schemes, although the basic idea was retained. For example, some of the classes were abandoned since they turned out to be mere duplications of other classes (due primarily to the poor quality of the original photos). But, even as old categories were jettisoned, new ones had to be created. In the final scheme, O was first, followed by B, A, F, G, K, and M. (Teachers quickly pounced on the memory aid: "Oh, Be A Fine Girl, Kiss Me Right Now.")

Pickering's work here was so widely heralded and publicized that astronomy soon captured the attention of the American people. Predictably, the rate of accumulation of data increased as more people entered the field and new technology emerged. Almost always with the aid of photography, astronomers have discovered thousands of new nebulas in the many decades since its introduction. And just as Darwin's pioneering work gave us an understanding of the origins of life that was nothing less than revolutionary, so had photography revealed new horizons never dreamed of before.

WOMEN IN ASTRONOMY

Interestingly, the burgeoning field of astronomy opened new career windows for women, although the actual numbers were comparatively small. The classification of spectra was both tedious and difficult. Once, one of Pickering's usually competent male assistants committed a careless error through just such boredom. Pickering, never known for tolerance or benign temperament, lost his patience and said, "Damn it, my cook can do a better job than that." And she did. Wilhelmina Fleming, cook-turned-astro-physicist, did unsurpassed work for several decades, even though she never received a degree in science.

It is customary to think of Maria Mitchell of Vassar when one thinks of the earliest female astronomers. In 1848, she was the first woman to become an honorary member of the American Academy of Sciences. She had discovered a comet, computed its orbit, and discovered several new nebulas. For this, the Danish king awarded her his country's prestigious gold medal. Shyness was not one of her weaknesses. As historian of science, Post points out, she was not afraid to opine on Bache's money-raising tactics. That took courage: Alexander Bache was highly influential, the first head of the National Academy of Sciences, created in 1863 by President Lincoln. Bache had also done important work in terrestrial magnetism. And, as historians Sinclair and Reingold note, he worked with the great British scientist Sir Edward Sabine of Edinburgh and helped give science its peculiarly American "flavor."

But as the case of Fleming shows, other women did superb work as well. Pickering had always been sensitive to the needs of women and, under his influence, many other women entered astronomy. Annie Jump Cannon, for example, catalogued over 200,000 stars. Cannon, who died in 1941, was curator of photographs at the Harvard Observatory for many years and as writer and historian Timothy Ferris says, " . . . though her sex barred her from all but the most grudging official recognition at the time, her work helped lay a foundation of modern astrophysics." Even today, astronomy is one of the branches of science in which women find it comparatively easy to gain entry, perhaps because of the tradition started with scientists like Vassar astronomers Maria Mitchell and Cannon.

THE PLANET BEYOND NEPTUNE AND
THE OTHER PICKERING

William Henry Pickering was born in Boston on February 15, 1858, and graduated from MIT in 1879. In addition to astronomy, his interests ranged from mountain climbing to odysseys in unexplored regions of the globe. Always fascinated by solar eclipses, he voyaged to such far-off places as Grenada, Chile, Colorado, and California.

In 1891, he set up an observatory in the city of Arequipa in northern Peru, under the auspices of Harvard. A few years later, he started the Flagstaff Observatory in Arizona, where, by 1908, he had constructed the first comprehensive map of the surface of the moon. Astute use of new methods and his own ingenuity in astronomy led to his discovery, in 1899, of Phoebe, the ninth satellite of Saturn.

Doubtless one of the most discussed phases in Pickering's career, however, was the quest for Pluto. By the time Pickering had become involved, numerous other scientists had already been seeking the "planet beyond Neptune." The first suggestion of such a planet had surfaced on November 17, 1834, when an enthusiastic amateur astronomer, the Rev. Thomas Hussey, noted a conversation between himself and Alexis Bouvard, a distinguished French astronomer. Hussey had suspected that the peculiar motion of Uranus—its motion inconsistent with theoretical predictions—could be due to the gravitational pull of an undiscovered planet. Bouvard's reply was:

> It had occurred to him, and some correspondence had taken place between astronomer Peter Hansen of the Gotha Observatory in Sweden and himself respecting it. Hansen's opinion was, that one disturbing body would not satisfy the phenomenon, but that he conjectured there were two planets beyond Uranus.[9]

Not surprisingly, there was considerable disagreement among astronomers, since no one had yet observed such a planet. In August 1850, American scientist and mathematician Benjamin Peirce, in an address to the American Academy of Arts and Sciences, was terse and frank: "There can be . . . no planet beyond Neptune. . . ."[10]

That was a staggering blow. Peirce's opinions were, in the eyes of some, just short of canonical. His son, C. S. Peirce, was a distinguished advocate of the pragmatist school of thought in

America—the view that concepts and ideas are meaningful only insofar as they have a practical application.

Benjamin Peirce had published a research paper while still an undergraduate at Harvard and earned his bachelor's in 1825. After receiving his Ph.D. in 1828 from the University of Leipzig, he spent a year studying with the great physicist Hermann von Helmholtz in Berlin. Also, with the cooperation of the Coast Survey, an official governmental branch working to further scientific research, he had been the first to successfully measure the curvature of the Earth. Despite Peirce's pronouncement, the believers in an undiscovered planet forged ahead. The renowned American historian of science Daniel Kevles, in his book *The Physicists*, points out that another important step in the search came on June 15, 1880, when David Peck Todd of the U.S. Nautical Almanac Office wrote to Professor George Forbes of the University of Glasgow. Forbes had long believed in planets farther out than Neptune, and even said, in 1880, that "There could be no longer a doubt that two planets exist beyond the orbit of Neptune. . . ."[11] In his 1880 letter, Todd wrote:

> Of course all my work was . . . inconclusive, as there are not, even up to the present moment, any well-marked residuals in the case of either Uranus or Neptune: . . . But, at the same time, I thought well enough of the work to attempt a practical search for a trans-Neptunian planet. It was conducted with the great refractor of the Naval Observatary [sic] during the latter part of 1877 and early in 1878.[12]

Even so, absolute confirmation of the new planet was still not forthcoming. Interest in it waxed and waned throughout the last two decades of the 19th century. For many, it had become a near-obsession, attracting both serious scientists and religious fanatics of all stripes, who saw the planet as a portent from the Almighty as the next century approached.

Around the turn of the century, Pickering finally became actively interested in the quest. (Edward Pickering, incidentally, had a hand in the search as well: using the very photographic techniques he'd helped develop, he first studied the data of the Danish astronomer Hans Lau, who had also believed that only two planets could account for the disturbances in the orbit of Uranus, but the search yielded nothing.) Pickering kept searching the heavens, taking innumerable photos from the Observatory in Arequipa, though the photos were inconclusive.

On November 30, 1908, Pickering published the results of his search in the *Harvard College Observatory* (circular No. 144), saying the evidence, consisting of calculations of disturbances or perturbations in the orbits of the planet Uranus, did indeed suggest there was an undiscovered planet farther out than Neptune, and he even gave the details of its position. Pickering communicated the results to the American Academy of Arts and Sciences on November 11, 1908. Because of his extensive work, numerous other investigators joined in the hunt. As is well known, the mysterious planet finally showed itself, making its debut on February 18, 1930, at the Lowell Observatory in Flagstaff, Arizona. This definitive discovery had taken so many years merely because the techniques of astronomical work were still in very primitive stages; much luck was required even to get a readable photograph. But by 1930, the techniques of astronomical photography were sufficiently advanced that the photographic plate clearly indicated a very distant planet beyond Neptune.

Pluto had been found. The conquerer of the elusive Pluto was Clyde Tombaugh of the Flagstaff Observatory. Still, much of the credit must go to Pickering, who wrote about and studied the issue perhaps more than did any other scientist: it often happens that a single person rather easily makes a find, but only because all of the preparatory theoretical and practical spadework has already been done by others. That this is a valid reason for spreading the credit around is all the more true when one considers that astronomical work was—and still is—difficult and, at times, even dangerous. Most observatories are located on high mountains, with the consequent unpredictable and often bitterly cold weather. The nature of the work required that the astronomer sit outside, in all kinds of weather, for hours, even days on end, as Pickering and others had done. The risk of exposure was high and the health of many scientists suffered.

THE SHAPLEY ERA

The Pickering era began to come to an end, however, when Edward Charles Pickering died of pneumonia in 1919 after directing the observatory for 43 years. (William Pickering lived on until 1938.) Under his guidance, it had become one of the world's great institutions, with numerous major research programs going on simultaneously. His loss was certainly a blow to the cause of women in astronomy and the scientific world generally. Of course, this created a dilemma. Who would be his successor? Almost immediately, suggestions started bouncing around the astrophysical community. On February 7, 1919, Dutch astronomer J. C. Kapteyn wrote:

> Today the news of Prof. Pickering's death reached us. It has been a great shock to us. Although our views in matters astronomical had little in common, I feel his loss severely. . . . My own conclusion, since some time, has been that his successor ought to be a student of statistical astronomy, and at the same time a good practical astronomer. . . . Shapley is a brilliant man and personally I, who knew him mainly through his scientific work, would think him best fitted for the position.[13]

Certainly, Kapteyn's criteria and recommendations had to be studied. He was already famous for his "statistical" studies of stars, which involved searching selected areas of the sky, gathering all data possible, and extrapolating to the rest of the sky. This procedure influenced virtually all future astronomers, including perhaps the most brilliant of them all, Edwin Hubble.

But despite Kapteyn's recommendation, Shapley's scientific skills were not sufficient to satisfy many members of the astronomical community, causing many sleepless nights for Harvard's President Lowell. One of the problems was Shapley's relative youth and inexperience in administration. For example, distinguished astronomer George Ellery Hale, the father of the great Hale telescope on Mt. Palomar, though he had initially supported Shapley, eventually withdrew that support, if only temporarily. He wrote to Lowell, saying that "While I am inclined to pick him as decidedly the most promising of all the younger group of American astronomers, I should be unwilling at present to turn over the directorship of this Observatory to him. . . ."[14]

A mysterious reply, given his earlier support. Shapley had, after all, already done pioneering work at Mt. Wilson, probing globular clusters—stars that appeared to be isolated, but were within the Milky Way. It was suggested that Shapley's political affinities may have alienated some. It is true that there is some evidence for this much later in his life. In a letter from President Hoover to Secretary of the Interior Fred Seaton, Hoover claimed, "This analysis reveals some of the efforts being made by communists . . . to exploit the 'fall-out' controversy. . . . Harlow Shapley and Edward U. Condon are among those creating fear . . . and confusion in the minds of the public." But Hoover was seeing "communists" everywhere. And even if Shapley may have leaned toward the left later in life, there is no evidence that I've been able to find for such leanings earlier in his career.

The plot was getting complicated. George Agassiz (grandson of the distinguished biologist Louis Agassiz) was shocked, saying, "Hale's behavior in this matter is to say the least peculiar. I can only account for it by the fact that at the last moment he is not willing to assume the responsibility of recommending so young a man."[15] Hale never did make his objections to Shapley any clearer, retreating to vague and borderline-contradictory comments. At one point, in a later letter to Lowell, he says, "I think that he would show great originality and ability as Director of the Harvard Observatory and that he would make very effective use of the Observatory staff and equipment."[16]

But in the very same letter he almost takes it all back by saying, "On the other hand, I clearly remember saying that in view of Shapley's daring and comparative youth, you would necessarily take some chances in offering him the place."

This seems insufficient: Why should Shapley's "youth" necessarily constitute a problem? Hale never tells us. In fact, Hale presents no evidence of anything specific about Shapley that would justify this caution. Harvard physicist and historian of science Owen Gingerich points out that Agassiz believed that Edwin Frost, then director of the Yerkes Observatory who had recently been at Mt. Wilson, may have influenced Hale, since Frost did not think Shapley was a good choice for director.

Lowell, finally, was rescued from this dilemma with the assistance of Julian Coolidge of the math department at Harvard. He hired Shapley after all. All in all, the choice turned out well enough. Among his other accomplishments, in 1918 he was the first to determine the rough dimensions of the Milky Way.

Yet whatever the truth of this controversy, the Harvard Observatory continued its unending series of contributions to physics and astronomy, all of it made possible in large part because of the efforts of the Pickerings, Shapley, and many others.

CHAPTER 2

Henry Rowland and Electricity

THE FIRST MODERN PHYSICIST

1883 was a year of commanding practical victories in science and technology: Robert Koch devised a method for inoculating people against anthrax, Edison was finishing his design of the first U.S. hydroelectric plant in Wisconsin, and British scientist Sir Joseph Swan produced the first synthetic fiber.

Yet, in a vice-presidential address to the American Association for the Advancement of Science (AAAS) that same year, Henry Rowland, one of America's first and most distinguished experimental physicists and mentor of the burgeoning Johns Hopkins graduate school in physics, made it clear that he was no follower of the Edison method: he was drawn to pure science, and not to gadgets like toasters and irons. He further declared that such fields as telephony, the study of motors, the telegraph, etc., should no longer be called "science." This was merely "technology," or the application of scientific ideas to practical, everyday concerns—an attitude revealing the escalation of scientific arrogance that would appear over and over again.

CHILDHOOD

Henry Augustus Rowland was born on November 27, 1848, in Honesdale, Pennsylvania. When he was 12 years old, he was already studying closely such distinguished journals as *Scientific American*. In 1868, scarcely out of his teen years, he wrote the following to his mother: "I intend to devote myself to science. If she gives me wealth, I will receive it as coming from a friend, but if not, I shall not murmur."[1] While science did not bring him fabulous wealth, he did become one of the great experimental physicists of the age.

EARLY CAREER

He graduated from Rensselaer Polytechnic Institute in 1870, returning there as an instructor two years later. Indeed, it was during his years at RPI that he first began serious work in physics both in class and in the darkened chambers of his boarding house.

Still, even though he was appointed to the faculty at RPI in 1872, he found the laboratory facilities disappointing. As he wrote to his mother in 1875, "I have been working for the last week in a little shed attached to the institute with the thermometer down so far that my breath froze on my instruments."[2] His career took a dramatic upward turn in 1875 when he received an offer from Daniel Coit Gilman, the president of Johns Hopkins, to serve as the first physics faculty member at that institution. It was a hard offer to spurn, for among its many other virtues, the university had, arguably, the finest collection of scientific apparatus in the world.[3]

Gilman himself is worth at least a passing comment: it was intensely difficult to ignore a personal offer from Gilman, a man of great influence and power. A good example of this was his willingness to dun Andrew Carnegie himself for money to aid the institution. After retiring from Johns Hopkins in 1901, Gilman became president of the Washington Memorial Institution, founded to assist scientists financially for worthwhile projects.

OFF TO EUROPE

Before teaching at Johns Hopkins, however, Rowland began to think seriously of heading for Europe to visit the major loci of scientific research. Europe was the place to study science and it was *de rigueur* for an American to study in Europe, both toward the end of the 19th and well into the 20th century. To underscore this point, it might be noted that John Trowbridge of Harvard, noted for his introduction of laboratory work in the training of physics in the 1870s, was almost the *only* American scientist of that era who did not study abroad, as reported by the contemporary historian E. H. Hall in his study of the physics department at Harvard. And Trowbridge was not among the superstars of American science in those days. (Trowbridge, in 1863, had served with the mathematician Benjamin Peirce on a commission of the Navy charged with armoring the steamer *Circassia* for wartime use.)

For a number of years, about the only one of Rowland's scientific contemporaries who grasped the full import of his work was James Clerk Maxwell. This was no small victory, for Maxwell had published his great treatise "Electricity and Magnetism" just a few years earlier and was already one of the world's most exalted physicists. So, in 1881, Rowland headed for London and Paris to demonstrate his new research. Maxwell's initial impressions of Rowland were overwhelmingly favorable. His contact with Maxwell was close, with Rowland even visiting him at his home. In the audience of one of his lectures, Maxwell was enchanted. And though Rowland's American prestige was limited, Europe soon ranked him among the world's elite.

So far as his attitude toward "tinkerers" was concerned, Rowland, though certainly unfair, did represent a trend in this epoch. To a great degree following Rowland's lead, increasing numbers of scientists began to feel that there was—or ought to be—a sharp distinction between those who dealt in "pure" science and the happy amateurs who indulged themselves by merely tinkering with gadgets in the backyard. (Ironically, Rowland, earlier in life, had himself shown a decidedly practical outlook. Debates rage even today in some circles; e.g., when "theoretical" physicists sneer at "experimental" work as a mere handmaiden or assistants to the flights of "real" intellectual handiwork wrought by the theorists.)

However, many scientists, including Simon Newcomb, one of the early important American astronomers, not only contributed enormously to "real" (i.e., theoretical, as opposed to mere experimental) science, but they and others were among the earliest to express their dismay at the poverty of mathematical instruction in the United States. They felt that such instruction was vital for both theoretical and experimental work. They also spoke out for the need for more scientific journals and societies, by which scientists might communicate their findings to one another. Similar-thinking scientists included the likes of the theoretical physicist and founder of the science of thermodynamics J. Willard Gibbs and Simon Newcomb. And, after the turn of the century, there surfaced men like Linus Pauling, Thomas Hunt Morgan, and, later, J. Robert Oppenheimer and Richard Feynman. The "tinkerer/amateur" epoch of American science would soon become history.

ROWLAND THE MAN

In many ways, Rowland was the very paradigm of the professorial stereotype in science. He is said to have bragged about the fact that he paid little attention to students, preferring his own research. Also, like most professors, he complained often about being underpaid. Unencumbered by trivia like modesty, he actually nominated himself for the Nobel Prize of 1901. The other American to do so the same year was R. H. Thurston, an engineer after whom the Mechanical Engineering building at Cornell was named.

He was equally pleased with his discovery of the laws governing magnetic circuits. A superb researcher, he produced, in his early years, one ground-breaking paper after another. Before long, he would, among many other things, measure the value of the mechanical equivalent of heat and produce a table of solar radiation frequencies with greater accuracy than anyone had done previously.

ROWLAND'S TEACHING

While Rowland was unquestionably a superb researcher, it seems that there is little room for debate about his teaching abilities. Soon after accepting the job at RPI, he became known as a miserable teacher. He thought a professor to be scarcely responsible for student learning. He felt that the acquisition of knowledge should be mainly due to a student's own efforts. Not only was this true at the graduate level (where such an attitude is somewhat more defensible), but he appeared to hold the same attitude toward undergraduates. And if he did not think a student had any real potential, he considered it a waste of time to spend any time working with the student.[4]

Some contemporaries did try to defend him, though rather unconvincingly. In the words of his colleague Joseph Ames, ". . . a student . . . might have received some help at various critical moments, but Rowland never regarded himself as responsible in any way for the investigation. . . ."[5]

Still, some students did fare well. As a matter of fact, by 1879, his first graduate student, William Jacques, did receive a Ph.D., though probably in spite of, rather than because of, Rowland's teaching. Jacques headed straight for the Bell Corporation, essentially as an electrical engineer—a field that Rowland unintentionally and indirectly helped found over a period of many years. There is, of course, a certain irony in this, as Rowland himself tended to disdain "tinkering" by contrast with "pure" science. Still, the fact that a body of engineering *theory* was beginning to emerge did tend to raise engineering above "mere" trial-and-error tinkering unguided by such theory.

ENGINEERING AND SCIENCE

Perhaps a bit more should be said about the place of engineering in the context of 19th century science. Rowland, as well as Michelson and Morley, the American physicists who conducted the famous "Michelson–Morley" experiment that is often claimed to have led to Einstein's theory of relativity (by proving that the notion of an "absolute" velocity of light was a senseless idea), were ably assisted by scientifically minded *engineers* of the day. Foremost among them was William Rogers, a fellow of the Academy of Arts and Sciences and a member of the AAAS.

In an essay on engineering in America,[6] historians Evans and Warner say, for example, that Rogers's engineering designs greatly influenced the designs of Michelson and Rowland. (In fact, Rogers identified an error in Rowland's work on engines.)

Interestingly, earlier in his life, Rowland had been trained as a civil engineer. But even in ostensibly "engineering" work, his work in pure physics proved to be more important. For example, it was mainly because of the reputation gained for his theory of electric currents rather than his engineering background that convinced the Niagara Cataract Construction Company to hire him to help them harness the power of Niagara Falls. So the Niagara event marked the beginning of a new era in American science. Physics, so long a poor stepchild of engineering, was coming into its own. Both, however, were and remain important theoretical disciplines.

ELECTROMAGNETISM

Another important area of Rowland's work that has not received as much attention as his later work, was his theoretical research on electromagnetism. It was this work that first attracted the attention of James Clerk Maxwell and also proved that America, even at this early period, had ". . . an experimental physicist who could pursue his profession at a level equal to the best of his European colleagues. . . ."[7]

His notebook of 1868 shows that his theoretical notions about electromagnetism were already incubating. Most of them were based on Faraday's *The Experimental Researches in Electricity*, though many of them were his own. Yet the ideas of Faraday figured so importantly that Rowland's own notebooks bore the title "Faraday's Experimental Researches in Electricity." The theoretical idea that stirred Rowland was Faraday's comment in the *Researches* that "If a ball be electrified positively in the middle of a room and then be moved in any direction, [magnetic] effects will be produced, as if a 'current' in the same direction . . . had existed."

The fundamental question was whether a magnetic field could be produced by a current alone, or whether it resulted from the interaction of a moving electrical charge *and* the metallic conductor in which the charge traveled. If the latter, then only electricity moving through, say, an iron bar would produce a magnetic field; a lightning bolt traveling through the atmosphere, for example, would not (unless it happened to strike a piece of iron). He stated his problem in an 1875 letter to Helmholtz:

> The question I first wish to take up is that of whether it is the mere motion of something through space which produces the magnetic effect of an electric current, or whether those effects are due to some change in the conducting body which, by affecting some medium around the body, produces the magnetic effect.[8]

Before these theories were tested in Berlin, Rowland was inclined to believe that a magnetic field was generated only in the conducting substance, and that an electric field alone would not produce magnetism. After the tests, he had to conclude that "electricity produces nearly if not quite the same magnetic effect in the case of convection [circulatory motion occurring in any area of uneven temperature, as when air circulates in an ordinary convection oven] as of conduction. . ." (the passage of electrons through a medium such as iron, copper, or other conducting substance).

Results like this eventually dissipated his faith in the old idea that electricity was a "fluid" that flowed along wires, although he apparently continued to believe the comparison of electricity to a fluid had been useful in some ways. In any case, by the mid-1880s, he began to view electricity more as a property of solid matter. It was just this point he emphasized at the 1884 physics conference in Philadelphia as well as to the Electrical Club in New York four years later.

In the course of these experiments he had also shown his great indebtedness to Faraday, by trying to exactly measure the direction and strength of the magnetic lines of force Faraday had described. However, since they varied with different kinds of metals, he had great trouble getting precise measurements. The reason was that the galvanometer readings fluctuated wildly and unpredictably as he altered the current in the magnets. The crudity of the equipment did not help either.

LAWS OF MAGNETISM

Still, one of the great outcomes of this work was that he managed to come up with a description of magnetism analogous to Ohm's law. In his careful and thoughtful article, "Rowland's magnetic analogy to Ohm's Law," historian David Miller describes this as follows:

> Thus in the center of the magnet the number of lines passing through the magnetic medium was proportional to M, the magnetizing force in the magnet, and inversely proportional to R, the resistance to these lines of force. He had therefore found a magnetic analogy to Ohm's Law for electric circuits. [In Ohm's law, the conductivity of a metal is inversely proportional to its resistance: the greater the resistance, the smaller the electric current.][9]

Soon, he stated a number of other laws governing electromagnetism. His most important contribution was doubtless his demonstration that what were ordinarily called electric currents were nothing more than moving electrically charged particles. Although this theory about the nature of electrical conductivity had been conjectured by a number of theorists previously, Rowland's research during this period constituted one of the first real, albeit still incomplete pieces of evidence for that theory.

STELLAR PHOTOGRAPHY

After his work on electricity and magnetism, Rowland became interested in solar spectroscopic analysis, the science that tries to determine the composition of the sun (or any other object) by studying the light emitted from it. Every element—hydrogen, iron, helium, etc.—has its own unique configuration of spectral lines, just as every person has his or her own unique fingerprints.

Before he tackled the solar work, however, he realized he needed more accurate diffraction gratings. Gratings in some ways acted like prisms, breaking ordinary light into its constituent colors. And although prisms had served well for earlier spectroscopic studies, he believed that he needed more precise ones for the kind of work he was planning.

Thus emerged his machine that would rule, with absolute precision, over 40,000 parallel lines per inch on concave glass and other surfaces. The gratings, breaking up visible light into its prismatic components more clearly than ever before, merely added to his growing eminence in the world of physics. He then began analyzing the spectra of all of the elements as accurately as possible. He published most of the results of this work between 1895 and 1900.

Soon, the over forty years of spectroscopic analysis would culminate in one of the most magnificent achievements in science, the discovery of the principles of quantum mechanics in the 1920s and 1930s.

Quantum mechanics represented a nearly complete break with the "classical" physics of Newton. Where all changes in nature occurred gradually for Newton, quantum mechanics showed that certain phenomena occur only in discrete "jumps"; for example, an electron will travel from one orbit to another in an atom, crossing a fixed distance and radiating a fixed amount of energy—the "quantum." Also disturbing to classical physics was the revolutionary idea that nature was intrinsically unpredictable. Where predictions in nature were concerned, no certainty could exist, only probable guesses.

This revolutionary work on the solar spectrum provided the foundation for later photographic work on the spectra of stars. He may not have created astrophotography, but with his new gratings he unquestionably did much to accelerate the progress in this field.

ROWLAND'S LATER YEARS

By the end of the 19th century, still other important changes had taken place. *The Physical Review*, one of the major journals of physics today, had been founded at Cornell University by Edward Nichols. When physicist William Anthony left Cornell in 1887 because the successor of Andrew White (the founder of Cornell), Charles Adams, was not sympathetic enough to science, he had recommended Edward Nichols as his successor. Soon to be a prime mover in American science, Anthony referred to him as " . . . the best man I know to make a success of the Physical Department . . . " (Nichols, a graduate of Cornell, had spent a year studying with Edison).[10] Though the journal published primarily Cornell faculty in its early stages, it became, within a few years, the equal of the best of the European journals. High and severe standards of peer review as well as better typesetting were soon introduced by Nichols.

Then, in 1899, Professor Arthur Webster of Clark University, a student of the fabled German physicist, Hermann von Helmholtz, founded the American Physical Society at Columbia University. As a fitting tribute, they elected Rowland, though fatally ill with diabetes, to be the first president of the society. Indeed, with these institutions, American science had taken another leap toward full professional status. In that same year, Reginald Fessenden would make history by transmitting clearly serviceable human speech via radio waves. And a few years later, in 1907, America would have its first Nobel Prize in A. A. Michelson for his work with the interferometer (a device consisting of various mirrors which was used by Michelson to measure the velocity of light).

Rowland had learned of his diabetes when he was married in 1890. Though the doctors estimated that he had about ten more years of life, his illness almost brought his laboratory work to a halt in the 1890s. But he persisted, continuing to discuss the nature of electrical conductivity with G. F. Fitzgerald. In response to Fitzgerald's inquiry in 1894, Rowland confessed that magnetic attraction appeared to act on the wires themselves, rather than on the charge moving through the wires. As he told Helmholtz also, "I was in hope of finding that the magnetic action of a current in the wire was due to some change in the wire caused by the current. This seems is not so. And yet the forces are between the wires! This is still a puzzle to me."[11] He continued to investigate the "ether"—the medium in which light was supposed to travel—and its alleged role in electromagnetic conductivity, though he had by now pretty much abandoned the "fluid" idea mentioned earlier.

Although Edison is the best known for telegraphic work, Rowland, later in his career, also created a system of telegraphy that allowed more than one message to be sent and received on the same line at the same time, which he called "multiplex" telegraphy. (Perhaps something of an inconsistency, given his attitude toward "tinkerers." But there is a difference between *experimental* science, based on solid scientific theory, and mere "invention." His telegraphic work arguably falls into the former category.) After his death in 1901, and the death of the great theoretician who developed the principles of thermodynamics, J. Willard Gibbs in 1903, there was something of a vacuum in physics and chemistry. For the next few decades, only a very few American scientists approached their level of accomplishments.

CHAPTER 3

J. Willard Gibbs and the Origins of Physical Chemistry

Josiah Willard Gibbs was the first distinguished American theoretician of physics and the first theoretical physical chemist. As testimony to this, on April 21, 1879, he was elected to membership in the National Academy of Sciences. This organization was a body that had emerged from the strife of the Civil War and was open only to the very finest scientific minds. In fact, Gibbs, at 40 years of age, was several years below the average age of its members and, at the time, could boast of only three published papers. But they were papers of surpassing importance. According to some of his colleagues, Gibbs felt that there were already enough people doing experiments, and what was needed was a theoretician to explain all of the accumulated data. And this he did superbly. His first accomplishment showed, using elegant diagrams, how the physical integrity of any substance (or mixture of substances) was affected by changes in such properties as volume and pressure.

CHILDHOOD

Gibbs was born in 1839, a contemporary of such luminaries as John Pierpont Morgan and Charles Dickens. But while these personalities shone in business and literature, Gibbs rose to prominence in other branches of human knowledge. Gibbs's father was a theologian. His lineage included eight Harvard graduates, three from Yale, and two from Princeton.

On his mother's side, Jonathan Dickinson became the first president of Princeton. On his father's side, Samuel Willard was pastor of the renowned Old South Church in Boston. Another relative became a secretary of the Massachusetts Colony.

When Gibbs was a child, the world was scientifically a long way from where it is today. Until age seven, Gibbs and his family lived in a house on Crown Street owned by the president of Yale, George E. Day. After Day retired, the family moved to 121 High Street in New Haven. (On a visit to New Haven, I noted that the house still stands, with a plaque indicating that it was the Gibbs family home.)

The times were changing rapidly. In the years after the second war with England in 1812, westward expansion was on the rise and industry was growing rapidly. Steam power was spreading and agricultural methods were improving. And in 1847, when Gibbs was eight years old, German physicist Julius Mayer had already stated the law of conservation of energy, noting that heat and mechanical energy are really two manifestations of the same fundamental phenomenon.

At age ten, his family sent him to a tiny, private school within a mile from their house in New Haven. Though not particularly gregarious, and even a bit shy, Gibbs led a fairly normal childhood. But he remained aloof to political events, a characteristic that diminished, but did not vanish in later life. Certainly he held religious convictions, but he had a long-standing aversion to expressing any opinions on anything he did not believe could be known with Aristotelian certainty.

The letters in the "Baldwin Papers" preserved at the Yale Library show that Gibbs certainly had an interest in girls. One unidentified writer, perhaps a jealous competitor for Gibbs's affection, describes Gibbs's relation with a girl named Fanny; Gibbs brought her "a most beautiful bunch of pond lilies . . . but Fanny does not like him at all, and we tease her about him, but she likes his flowers much better than she likes him."[1]

COLLEGE YEARS

A critical moment in his formative years came at the moment he was to take his entrance examination for Yale in 1854. He became paralyzed with self-doubt, dreading the possibility of not passing the entrance examination. In the 1850s, oral exams were still the rule. In an article titled "One of the Prophets," the distinguished historian Margaret Whitney describes Gibbs's oral exam:

> Edith Woolsey has a story that she thinks came from her father, that being very shy he dreaded the entrance examinations and went to his good friend [Yale] Prof. Thacher. . . . Prof. Thacher questioned him on a variety of subjects and [told him] that he had passed the equivalent of an entrance examination and was admitted to college.[2]

Thacher then supposedly kept trying to calm him and build some measure of self-confidence. How much weight one can put on such stories is arguable, but that he performed radiantly at Yale is a fact. In his junior year, for example, he earned the Berkeley prize for Latin.

TOWN AND GOWN

Of course, Yale manifested the usual college mayhem. Some student pranks so infuriated the people of New Haven that town–gown relations reached their lowest ebb in many decades during

Gibbs's undergraduate days. Both before and after the Civil War, students nationwide had created secret organizations with a bizarre variety of initiation practices. Yale was no exception. One in particular dramatizes the tensions of this epoch.

During Gibbs's first year at Yale, two citizens of New Haven died during a particularly brutal attack by a gang of students— some even carrying firearms. So high did emotions soar that during a similar incident, the town, in its unimaginable fury, took over the town cannon and tried to turn it on the students. At best, the most tolerant citizens of New Haven could see little point and much evil in such mayhem. Soon, a virtual state of siege existed between college and town.

There is no evidence that Gibbs ever took part in these events. He did join in most "normal" student activities and was a member of "Brothers in Unity" (something like a modern fraternity, but emphasizing debating) but the records show that he was hardly an active participant there either. Again, the Baldwin Papers at Yale show that his chief interest lay in taking long, solitary walks, a habit he continued throughout his life.

GRADUATE SCHOOL

Gibbs had already begun thinking about graduate school during his last years at Yale. But where? Much of America was vast, unexplored territory, and education, even at schools like Yale, tended to offer practical training, intended to equip people to deal with the changing frontiers. And apparently, Gibbs's early inclinations at least, were practical, despite the fact that his great contributions to science and his passport to immortality were in theoretical chemistry and physics.

In any case, another institution appeared in these antebellum days, called the Sheffield Scientific School in New Haven. This was a division of Yale University honoring Joseph Earl Sheffield, who used his fortune to create the first endowment as well as the first buildings and equipment. Despite its having had provisions for graduate study since 1847, economic and political problems conspired to prevent the initiation of its graduate program for several years.

The school did not award a Ph.D. until 1861. Yet it had a rich history. In the years just before and including the offering of its first Ph.D., the Sheffield school had cooperated with the famed California Geological Survey, allowing some of its scientists to use their facilities to study ore samples. By the 1880s it was becoming more of a graduate school devoted to "pure" science rather than an engineering school. To some extent, the interests of the students it was attracting brought this about. Doubtless too, many of the faculty were interested more in "pure" science than in engineering.

So it was, in June 1863, that Gibbs became the first American ever to be awarded a Ph.D. in engineering, for a dissertation titled "On the Form of the Teeth of Wheels in Spur Gearing." Gibbs was one of only three to receive it that year in science. (Yale had, in fact, been the first school in the United States to award that degree, and the year it gave the award to Gibbs was the third year that Yale conferred the Ph.D. in any field.)

It's difficult to be sure who his teachers and mentors were during his graduate years, although the Yale catalogue indicates that he probably took courses with W. A. Norton and Chester Lyman in engineering, as well as the renowned Benjamin Silliman, Jr. in chemistry. Gibbs had entered Yale's Department of Philosophy and the Arts in September 1858 as an engineering student. In 1861, Yale renamed the department the Sheffield Scientific School.

OFF TO EUROPE

As was common in those days (though less so later on), the college invited him to stay on the faculty to tutor undergraduates. (At the time, Yale made the assumption that any graduate made a tutor must be competent to teach any of the first courses he had taken in the freshman and sophomore years.) In 1862, an inheritance had freed Gibbs from financial worries at least for a while. In that year, Gibbs and his siblings acquired their father's estate, roughly $23,500—a fortune for that time.

After the term of his tutorial at Yale expired, Gibbs, like most of the scientists of the day, finally made the trek to Europe to study the latest advances in chemistry and physics. Soon he was poring over the research of such luminaries as Joseph Lagrange, S. D. Poisson, and Pierre Laplace. After several months, feeling he had exhausted this fount, he went to Heidelberg to attend similar classes given by the likes of Georg Cantor, Gustav Kirchhoff, Hermann von Helmholtz, and Robert Bunsen. Gibbs began his work on thermodynamics by studying the laws of thermodynamics as given by Rudolf Clausius, the 19th century German physicist at the University of Zurich. Clausius was the first to really put thermodynamics squarely within the scientific universe. He is best known for stating in elegant mathematical form the principle that heat cannot pass from a hotter to a colder body.

In the 1870s, Gibbs described the mathematical conditions for chemical equilibrium. In his view, "equilibrium" referred to any substance equally balanced between two or more states. (An example would be a closed system where water is changing into steam as fast as the steam is changing into water.) He then showed how such systems would be further affected by changes in gravitational forces, the addition of other chemical elements, external pressure, surface tensions, and so forth.

A further milestone in the development of American physical chemistry, a branch of chemistry dealing with physical properties of substances and how they relate to their chemical composition, dates back to 1897. It was in that year that Professor Wilder Bancroft and James Trevor founded the *Journal of Physical Chemistry* at Cornell University. Gibbs, in turn, midwifed the birth of this journal.

In an article written much later, George Uhlenbeck, renowned for his discovery of the spinning electron, paid great testimony to Gibbs:

It was hard going; I had to learn analytical mechanics and several branches of mathematics just to be able to follow the argument. . . . I also dipped into Gibbs' *Statistical Mechanics* [Gibbs' opus magnum, which explained the behavior of large numbers of particles using ideas based on probability theory]. . . ."[3]

In a similar vein, Heinrich Hertz, the 19th century German physicist, after whom radio waves are named, writes:

Dear Sir:

    I give you my best thanks for the papers you were so kind to send me. I beg [you] to accept in return those of my electrical papers of which I have copies left. It is with great pleasure that I seize the opportunity to give expression to the very highest esteem I have since long had for your thermodynamical work. I did not master it absolutely but I mastered it enough to see how fundamental it is. Many things which I thought had been first done by Helmholtz [a 19th century German physicist who explored the connections between physics and biology] I found in your paper. . . . If it took much time that your paper was fully appreciated here I think it is because the Cambridge(!) Transactions are only with difficulty to be had here. . . .[4]

These letters show the high regard Gibbs had in Europe—if not yet in America.

    Similar praise came from the likes of J. H. van der Waals, the Dutch physicist after whom the "van der Waals" forces (important in explaining the structure of protein molecules) are named; James Clerk Maxwell, the legendary 19th century British physicist and contributor to the theory of electricity and magnetism; and many other luminaries of the day. Gibbs had not yet reached either Hertz's or Maxwell's already legendary status. But this gap would not last for long.

    Maxwell's appreciation especially turned out to be of great practical importance. Indeed, Maxwell, in his famous "Nineteenth Century Clouds over the Dynamical Theory of Heat and Light," shows that he too was interested in issues bearing on Gibbs's work. Of the many letters in the Gibbs Collection at Yale University, twelve are from students or friends of Maxwell. (Whether Maxwell himself ever wrote to Gibbs is not clear.) It is a matter of historical record that Maxwell brought Gibbs's work to the attention of his own students as well as many other American scientists. Maxwell was then director of the prestigious Cavendish Laboratory of Experimental Physics in London, an appointment given only to the very greatest British scientists. He had been following American physics assiduously, and being a generous man, never hesitated to help a promising young star.

In contrast, for years Henry Rowland had been trying to get his work recognized, but with only halting success. When he penned a paper on ferromagnetism, for example, the *American Journal of Science* rejected it, believing it contained nothing particularly new or unique. That was a blow, for this journal, founded in 1818 and edited by Benjamin Silliman, was arguably the most influential and important in American science. (Perhaps its only competitor for a while was the *Boston Journal of Philosophy and the Arts*.)

THE INACCESSIBILITY OF GIBBS

One of Gibbs's problems, certainly, was the murkiness of his writing. Also, since Gibbs was a theoretical physicist, he was inclined to express his ideas in a purely mathematical form, further adding to their inaccessibility. In the following letter, Cambridge scientist M. M. Pattison Muir writes of the difficulty of Gibbs's work as follows:

> Sir:
> . . . I was told of your paper on the "Equilibrium of Heterogeneous Substances" but being no mathematician . . . I did not venture to read it.—But recently I found Clerk Maxwell's translation of your paper into ordinary language, in the Science Conference at South Kensington (1876). Will you allow me as a chemist to thank you most sincerely for the very wonderful work you have done for us. . . . Your results seem to throw light on very dark spots in chemical theory. . . .[5]

But even if Gibbs had been more inclined to more conventional prose, it was a fact that communication generally between the continent and the United States was poor at the time. Not many Americans could afford or get hold of the European journals. Their books were hard to come by as well. Nor was there the plethora of scientific meetings, easily reached by airplane today.

So, foreign translations of his work scarcely helped his cause; consequently, he remained almost an unknown quantity in the United States. Some fault was his own, however. Unlike many scientists, Gibbs did not hunger after fame and was, therefore, unconcerned with publishing his work. In a similar vein, he preferred to stay with a very close circle of friends. These factors were such a hindrance that even after he had garnered the support of Maxwell, who promptly persuaded the *Philosophical Magazine* to publish his work, Gibbs's fame still grew slowly, even in Europe.

In fact, the only thing Gibbs did do to circulate his work was to periodically distribute his papers to people on his private mailing list. In fact, the Yale records show that his mailing list included countries from Canada to Poland and India. For example, he sent his paper "On Multiple Algebra" to 276 people worldwide, which comprised a small group of scientists who Gibbs felt could appreciate his research.

Others had helped Gibbs as well, including American scientist Henry A. Rowland. Gibbs was understandably grateful to Rowland for this support. It was through Rowland that Gibbs was able to give a series of lectures on mechanics at Johns Hopkins, during January and February 1880. It was also through Rowland that Gibbs accepted one of the very few committee appointments in his life, as commissioner to the National Conference of Electricians at Philadelphia.

In a letter of March 3, 1879, Rowland (then at Johns Hopkins) wrote Gibbs (now and forever at Yale) as follows:

> I have also to thank you for sending me some time since your papers on Thermodynamics. . . . Mathematical physics is so little cultivated in this country and the style of work is in general so superficial that we are proud to have at least one in the country who can uphold its honor in that direction.[6]

Everyone was hardly less impressed with Gibbs's two papers, "Graphical Methods in the Thermodynamics of Fluids" and "A

Method of Geometrical Representation of the Thermodynamic Properties of Substance by Means of Surfaces." These papers convinced Maxwell that Gibbs was a genius of the first order, capable of ground-breaking work in science. In fact, Maxwell had announced to English chemists at a meeting of the Chemical Society of London earlier in the same year, 1879, that because of Gibbs's work, "problems which have long resisted the efforts of myself and others may be solved at once."[7]

Unfortunately for Gibbs, Maxwell died not long afterwards at the premature age of 47. Gibbs's work then remained buried until the influential German Friedrich Ostwald, a pioneer in the study of ionization, took over where Maxwell left off. He saw to it that a German version of Gibbs's work appeared in print—a not unimportant fact, for as noted earlier, most European scientists regarded America, rightly or wrongly, as a nation of amateur tinkerers at worst, and competent experimenters at best. In his autobiography, Ostwald describes Gibbs's thermodynamic work as "awesome" and says that "it influenced [my] own scientific progress considerably. . .":

> While still in Dorpat, Oettingen had brought my attention to the American J. Willard Gibbs's thermodynamic writings as being unusually important, if difficult investigations. . . . Like Oettingen, I found them difficult but saw that they were of vast importance. Before myself only a few . . . understood their significance. . . . His work influenced me. . . . Gibbs deals almost always with the parameters of energy and makes no assumptions about kinetics. Due to this, his work is on a high plane of certainty and a very high degree of lasting quality.[8]

GIBBS'S TEACHING

Evidently, he did not do well teaching undergraduates and some rumors even held that he almost lost his job because of his ineptness. During his 32 years of teaching he taught almost 100 *graduate* students. Historian Daniel Kevles has argued forcefully that Gibbs did not like having graduate students and even with the ones he did have, he shared very little of the details of his work.[9] Nonetheless, many of them went on to careers in teaching, medicine, and industry and made at least some minor contributions to science. Many of these former graduate students spoke extremely well of his lectures. One named Hastings, about whom little is known except that he was a student of Gibbs's for a number of years, says in his memoirs:

> No qualities of Professor Gibbs impressed his sympathetic associates and his pupils more than his serenity and apparent unconsciousness of his intellectual eminence. . . .

Most agreed, however, that it did require constant attention to follow Gibbs in class. Often, the students had to interrupt him to ask for clarification.

Doubtless some, undergraduates especially, were distracted by his mannerisms, such as leaning from side to side while he spoke and gazing out the window instead of at the class. One possibility is that his boyhood shyness reasserted itself in the classroom. Also, most critics of Gibbs's teaching overlook the fact that most of the negative reports about it came from his years when he was a young *tutor* at Yale—before he assumed a full professorial post.

However, in fairness to Gibbs's critics, it seems undeniable that he was preoccupied with his own research— not the first time that this failing has existed in the academic world. It was as true in those days as it is today that research and publication is really what matters for academic survival. After all is said and done, most schools care little about "good teaching."

Finally, other universities began to notice Gibbs. When it appeared that Hopkins might be luring him away from Yale, American geologist James Dana expressed the feelings of the Yale faculty in a letter of April 26, 1880:

> My dear Prof. Gibbs:
>   I have only just now learned that there is danger of your leaving us. Your departure would be a very bad one for Yale. . . . I do not wonder that Johns Hopkins wants your name and services, or that

you feel inclined to consider favorably their proposition, for nothing
has been done toward endowing your professorship and there are not
here the means or signs of progress which tend to incite courage in
Professors. . . . But I hope nevertheless that you will stand by us,
and that something will speedily be done by way of endowment to
show you that your services are really valued. . . .[10]

As history records, Gibbs remained. Yale, after all, had produced
him and he was accustomed to New Haven. Yet life was becoming
increasingly difficult for him. He had an invalid sister to look after.
By this time, a thin, bearded Willard Gibbs was over 40. As
always, he dressed impeccably. He had already been at Yale for
nine years, as professor of theoretical physics.

Another characteristic of Gibbs kept him at Yale. Though he
never married, he was accustomed to his domestic routine. The
paradigmatic academic that he was notwithstanding, he had
something of a household flair. He chipped in with all of the
household chores, including washing dishes, fixing toys, taking
his sister's children skating, and bandaging their bruises. Later,
they would of course know of his legendary status in science. But
at present he was just their uncle.

PHASE RULES

Gibbs understood that change is the essence of nature, as
Aristotle and Heraclitus realized, but Plato did not. Today, Gibbs is
rightly given credit for the founding of the branch of science that
explains such change—physical chemistry. It is one example,
along with the later application of quantum mechanics to chemistry by Linus Pauling, where physics and chemistry meet.

Gibbs was interested for many years in the concept of "entropy." That, in turn, was part and parcel of his interest in the
growing field of thermodynamics, the branch of science that
studies energy. Examples of the importance of thermodynamics
are not hard to find. Gibbs knew, for example, that if one heated a
liquid in a closed container, the liquid would gradually vaporize,

increasing the pressure of the gas against the container walls. This pressure, in turn, could do things like drive the piston of an automobile.

The explanation for this increased pressure runs as follows: the greater a molecule's energy, the higher will be its temperature, and the higher its temperature, the greater the velocity of its constituent particles. With the great velocity comes the great pressure the gas (such as steam) exerts and the more work it does.

Eventually, Gibbs's analysis of things like pressure, volume, and so forth resulted in his now well-known "phase rules." Basically, they described the "equilibrium" conditions for any substance. ["Equilibrium" is merely a kind of stability: a liquid substance, such as water, is in equilibrium with its gaseous phase when the gas (steam) is changing back into water just as fast as the water is changing into steam.] And Gibbs scarcely hesitated to offer these "phase rules" to the world in numerous pages of complex mathematics.

Before Gibbs's time, the only way to produce equilibria of the above type was by trial and error. For example, in order to produce an equilibrium of some combination of chemicals, the researcher simply had to keep changing the temperature, the container, and so forth, until he chanced upon the right combination. But with the phase rules the careful researcher could find out ahead of time precisely what conditions (e.g., temperature, pressure, proportions of chemicals used) would be needed to produce an equilibrium.

Among the many distinguished scientists Gibbs influenced was Johannes Diderik van der Waals. Van der Waals, in turn, brought this work to the attention of the Dutch chemist Hendrik Roozeboom, who had been having difficulties solving some problems in the manufacture of steel. Roozeboom knew that steel could be correctly produced only when the percentages of iron and carbon, as well as temperature and pressure conditions were just right. But what were these conditions?

Roozeboom's dilemma is a perfect example of the frustration faced by scientists before the phase rules came into being. He had been trying one combination after another without complete success, but once he had hold of the phase rules, he solved the problem immediately. (Even cooking has its "phase rules." Try making a soufflé without knowing the exact proportions of flour, the right oven temperature, etc.)

A similar problem existed with ammonia. A difficulty in making ammonia on a large, commercial scale was that the gas tended to decompose quickly into its constituent elements, nitrogen and hydrogen. The problem was important, since ammonia was an ingredient in the production of explosives as well as fertilizer. Later, in the early 1900s, German chemist Fritz Haber showed that this decomposition could be slowed by the use of extremely high pressures using the phase rules.

In the many years following Gibbs's work, many other scientists described in far more detail the applications of his ideas, enumerating applications of the phase rules to almost every science, including geology, physiology, astronomy, solid-state physics, biophysics, and many others. And these rules were later applied to mineralogy, the study of volcanoes, the formation of salt deposits on the ocean floor, the manufacture of new metal alloys, and even the production of portland cement, a type of cement formed by heating a mixture of limestone and clay.

Gibbs, as a theoretician, did not concern himself with practical applications. Instead, he went on extending and perfecting the theory of phase transitions. Eventually he was able to analyze phase changes not only in simple systems (where one substance, such as water, changes to steam) but complicated ones where several substances are all undergoing phase changes simultaneously.

THE SECOND LAW OF THERMODYNAMICS

Much work was done in the 19th century on the laws of thermodynamics. In 1824, French scientist Nicolas Carnot had studied the principles of the steam engine and thereby paved the way for both the first and second laws of thermodynamics. In

1850, German physicist Rudolf Clausius explicitly formulated the second law.

The first law is the principle of conservation of energy, meaning that you get as much energy out of a reaction as you put into it. The second law of thermodynamics says that the disorder ("entropy") in the universe can only increase, never decrease. In effect, entropy is a measure of disorder. And it can be measured in any energy transformation or chemical reaction. (A good example of entropy can be seen by comparing a college student's dorm room at the end of a semester with the way it looked at the beginning of the semester.)

It is interesting to note that a powerful conceptual bind existed between Gibbs's work and the work of his elder contemporary, the great Max Planck, who spent more than the first decade of his career in the study of the inevitability of decay as implied by the second law. (But Gibbs had understood the second law before Planck did.) Perhaps oddly, even after scientists had formulated and understood the second law of thermodynamics, practical, industrial applications did not develop for some time. Part of the trouble was that thermodynamics generally required a vast amount of profound mathematical knowledge. Gibbs had it, but not too many others did.

VECTOR ANALYSIS

Another, equally important contribution of Gibbs was "vector algebra." The idea of a vector is simple enough. In physics, *direction* is as important as pure numerical values. (If you are standing on railroad tracks, it makes some difference if the speeding train is heading *for* you or *away* from you.) So, a vector is any quantity possessing both a magnitude *and* a direction, and vector algebra is concerned with the directions of forces in nature as well as their magnitude.

He also showed the logical relationship between this idea and the second law of thermodynamics, inasmuch as the second, or "entropy" law states that natural processes are intrinsically directional. The second law states that energy goes naturally from a state of lower entropy to a state of higher entropy. (The student's room will not "reorder" itself if left untouched, but will just get worse over time.)

Gibbs's studies on vector algebra began with Maxwell's classic Treatise on Electricity and Magnetism, which first discussed and put Faraday's ideas on electrical currents in mathematical terms. In an 1888 letter to Dr. Victor Schlegel, another scientist also interested in the theoretical aspects of phase rules, he says:

> My dear Dr. Schlegel, I am glad to hear that you are pleased with my address on Multiple Algebra. . . . My first acquaintance with quaternions was on reading Maxwell's E. & M., where the Quaternion notations are considerably used. I became convinced that to master those subjects, it was necessary for mastering those methods. . . .[11]

It is not necessary to discuss "quaternions" here: suffice it to say that before Gibbs's "vector algebra," quaternions was virtually the only mathematical system that even began to do the things the vector algebra would do. But it was clumsy and inefficient and the letter suggests that Gibbs was not exactly overjoyed with it.

In the mathematical terms of Gibbs's system, every direction in three-dimensional space can be indicated by three coordinates. In this way, Gibbs was able to describe the equilibrium state of any thermodynamic process *pictorially* (somewhat suggestive of the way Richard Feynman later used "Feynman diagrams"). In Gibbs's algebra, the equilibrium of a thermodynamic system could be represented by a point with respect to a "surface." If the point was above the surface, the equilibrium was stable; if below, the equilibrium was unstable. In this way, the picture depicts the various equilibrium and disequilibrium possibilities, as pressure and temperature change.

It is interesting to note too that the topic of vectors led to the creation of one of the very few theoretical "textbooks" written in the United States before the turn of the century. This was actually a short pamphlet Gibbs wrote entitled *Elements of Vector Analysis*

Arranged for the Use of Students in Physics. Only a few others existed, e.g., *The Theory of Electricity and Magnetism* by A. G. Webster and *Elements of the Theory of the Newtonian Potential Function* by Harvard's B. O. Peirce.

THE PLAYFUL SIDE—CATS

Scientists have their playful side. Da Vinci studied birds to understand flight and Gibbs studied cats to understand the physics of falling objects. This phase of his work was of less than cataclysmic importance (cat lovers, presumably, will disagree). Gibbs even sought to understand why cats land on their feet. On December 4, 1894, he gave a talk titled "On Motions by Which Falling Animals May Be Able to Fall on Their Feet." According to his explanation, this was possible because of the cat's ability to alter its "moment of momentum." More simply, a cat can twist its body so that the center of gravity is below the midpoint of the body. Imagine holding a child's toy top upside down: if released, the heavy end naturally falls downward. As Gibbs showed, a cat can twist in such a way that it makes itself like an upside-down top, so, like the top, its feet are on the "heavy" end and they therefore swing downward so that the cat lands on its feet.

THE DARK SIDE

The dark side of Gibbs's work showed up dramatically during the second decade of the 20th century. In World War I, the German Fritz Haber (whose name is immortalized in what is now called the "Haber cycle," a process Haber invented which allowed industrialists to produce ammonia and nitrogen quickly and inexpensively, thereby revolutionizing agriculture), and Freeth, an Englishman, applied the phase rules in an apocalyptic way. Both had studied them thoroughly and were now able to solve critically important problems in manufacturing explosives. They accomplished this by studying the phase changes in nitrogenous compounds—compounds containing nitrogen, such as TNT.

A typical glitch in making bombs was that many nitrogenous compounds would decompose before they could ever be effectively used as explosives. But by applying the phase rules, Haber managed to solve the problem and the Germans had a far greater supply of nitroglycerin to use against the Allies than they would have had without Gibbs's work.

STATISTICAL MECHANICS AND THE FROZEN UNIVERSE

The work of German physicist Ludwig Boltzmann of the University of Vienna on statistical mechanics profoundly influenced Gibbs in his later years. Boltzmann began to view mechanics in terms of entire systems of particles, rather than focusing on the statistics of individual particles. (The idea of "statistical mechanics" is really fairly simple. Consider several thousand people in a fairground. Then suppose someone yells "Fire!" How can you predict what the crowd will do? One could study the behavior, history, and psychological makeup of each person and then generalize. It's easier, however, to make some assumption about what the statistically "average" person would do and make a probability judgment based on that. In our example, most of the crowd would likely run for the nearest exit. This may be a statistical "guess" and therefore probabilistic, rather than certain. But it's probably fairly sound. Statistical mechanics treats all of the atoms and molecules in a piece of matter in the same way.)

Gibbs fused this point of view with his own notion that the dynamics of statistical mechanics could be studied independently of the internal *structure* of substances. He says in his book *Elementary Principles of Statistical Mechanics*:

> The laws of thermodynamics . . . express the approximate and probable behavior of systems of a great number of particles, or, more precisely, they express the laws of mechanics for such systems. . . .

Moreover, we avoid the gravest difficulties when, giving up the attempt to frame hypotheses concerning the constitution of material bodies, we pursue statistical inquiries as a branch of rational mechanics.

In a nutshell, Gibbs is saying that to understand and make predictions about how systems of particles behave, it isn't necessary to understand the internal composition of that substance. By analogy, to understand the principle of flight, it isn't necessary to understand what an airplane is made of. The principles of flight are the same, whether the plane is made of aluminum, tin, balsa wood, or fabric and wire.

Here, Gibbs is already anticipating the coming revolution against classical ideas in physics, such as Einstein–Bose statistics and quantum mechanics. The common element between quantum mechanics and Gibbs's work is the vast importance of the element of probability. Quantum mechanics says the behavior of particles can only be approximately predicted and Gibbs also relies on similar statistical notions. (Of course, this is a stretch, since the idea of probability in quantum mechanics and Gibbs's work is applied very differently.) Still, it is unfortunate that Gibbs, unlike, say, Michelson, did not live to see the emergence of the "new physics."

In discussing the motion of subatomic particles using Newtonian concepts, such as Newton's law of action–reaction, which says that for every action there is an equal and opposite reaction, he further developed the theory behind statistical mechanics, by showing that a scientist could predict the observable properties of a substance by studying the behavior of the particles composing that substance. (The total energy of a bottle of gas, for example, could be predicted by studying the behavior of the molecules of that gas.) And his work had interesting implications for the future of the cosmos, which would later worry the great discoverer of the charge on the electron, Robert Millikan. Working with the old classical second law of thermodynamics, many of Gibbs's fellow scientists were foretelling the collapse and end of the universe. (The second law was not Gibbs's work—it was formulated by

French physicist B. P. Clapeyron in 1834 and German physicist Rudolf Clausius of the University of Zurich in 1850.) According to the second law of thermodynamics, entropy would eventually reach a maximum, i.e., the universe would reach a point where it would no longer be possible to use any of the energy in it. The cosmos would be cold, flat, and dead. The moment would be the "heat death," so called because in that final state, the universe would be completely disordered (its entropy at a maximum) and there would be no internal energy left to do any work, which H. G. Wells described so vividly and memorably in his novel *The Time Machine*.

Gibbs, however, with his work on statistical mechanics, argued that this dreaded possibility was more fiction than fact and that chances for survival of the cosmos were higher than others had thought. This "heat death" would probably never occur.

FINAL YEARS

In his final years, Gibbs had gone gray, often appearing grave and intent. As always, he remained polite and friendly to those who would study with the master, always patient and willing to go slowly with them. Although signs of age were appearing, including a receding hairline, in most ways he appeared no less robust than in earlier years. He continued to visit Vermont, enjoy classical music, and take long walks in the country. He remained less visible on campus than were many of the other faculty, refusing to the end to participate in the small talk at the Graduate Club.

He became seriously ill with a bladder infection in 1902, in the midst of preparing for the reprint of some of his work by Longmans, Green and Company publisher. Though he appeared to improve for a brief period, he became violently ill on April 26. At 11 o'clock, on April 28, 1903, he passed into history, with only a nephew and brother-in-law present. (His younger sister was in Europe at the time.) The family held the funeral at 121 High Street and buried him at Grove Street Cemetery.

A measure of Gibbs's importance in the history of American science as well as science generally, was the inauguration of the Willard Gibbs Award, founded by chemist W. A. Converse and sponsored by the Chicago section of the American Chemical Society. The society gives the medal every year to a chemist of outstanding abilities and contributions, and the list of recipients includes such luminaries as Linus Pauling.

Compared to some scientists, Gibbs did not write much during his lifetime. But what he did write, for worse or better, forever changed history.

CHAPTER 4

A. A. Michelson and the Ether

> While it is never safe to affirm that the future
> of Physical Science has no new marvels
> in store . . . it seems probable that most
> of the grand underlying principles have
> been firmly established, and that future
> advances are to be sought chiefly in the
> rigorous applications of those princi-
> ples. . . . It is here that the science of
> measurement shows its importance. . . ."[1]

With this remark, Nobel Prize-winning physicist A. A. Michelson
summed up his philosophy of science. The comment was delivered at the dedication of the University of Chicago's Ryerson
laboratory, named in honor of the legendary Illinois lumber baron
whose son, Martin Ryerson, donated $150,000 for a new lab.
Radically wrong as it proved to be, it demonstrates the colossal
egocentricity of one of the towering figures in the history of
American science. The year was 1898. The United States would
declare war on Spain over Cuba and Marie and Pierre Curie would
discover radium.

   Albert Abraham Michelson crystallized the attitude of
scientists—his own included—that dominated much of 19th century thinking about science. For Michelson personally, that

attitude had served him well. He was the first American ever to capture the Nobel Prize in physics for his precision optics and the spectroscopic and metrological investigations conducted through the use of these optics. Most importantly, Michelson would disprove the "ether" theory, the idea that light waves are carried through space by this mysterious, invisible substance called the ether.

Some historians, like Elizabeth Crawford in her excellent article on Nobel prizes, argue convincingly that his selection was in part due to a strong bias in favor of measurement projects in physics by the Uppsala University physics department, which had a powerful voice on the Nobel Committee for physics.[2] Apparently a bias toward "experimental" physics was not a purely American phenomenon as many historians of science have supposed.

EARLY LIFE

Michelson was born in the town of Strezelno in Poland in 1852, during a time of tremendous political upheaval which had begun in Europe four years earlier. During this difficult period, many sought refuge in America, including engineer Henry Flad and biologist Charles Mohr.

Soon the family settled in Virginia City, Nevada. The legendary Comstock Lode, a rich deposit of gold and silver discovered in Nevada in 1858, was located there and Michelson senior soon began his quest for quick money, a malady that had already afflicted thousands of Americans. Many pursued this dream both for personal profit and to dig out enough gold to help finance whatever side they supported in the raging Civil War. Almost immediately, his early optimism proved ill-founded and he lost nearly everything in his fruitless prospecting. That will-o'-the-wisp exorcised, he turned to more realistic business ventures and set up a general store to keep his family fed.

EARLY SCHOOLING

Fortunately, Michelson's parents did not neglect their son's education throughout their financial struggles, and Michelson began his academic life by attending a local grammar school where his parents encouraged him to study the violin and even billiards. Because of the reputation of San Francisco's first-rate schools, his parents decided to send him there. The nearer Virginia City schools were known to be singularly inferior. Fortunately, he ended up living in his science teacher's boarding house, which gave him a head start in science and an understanding of the scientific method, however rudimentary. It was not long before he and his family realized that science was Michelson's specialty. One avenue to that goal was medical school, something his mother preferred over any other career. The fly in the preparation was that the best medical schools were on the very distant East Coast and far too expensive for the family budget.

Eventually, his father decided that Annapolis would be the logical choice, since it offered an unusually strong college-level program in science. Although Michelson did not get in the first time and had to jump through a few political hoops to get in later on, no one could ignore his talent forever.

The Naval Academy had beefed up its scientific offerings considerably in those years. By the mid-1870s, it had a wider variety of as well as more advanced physics courses and math offerings. The spring 1870 finals, for instance, included considerably more difficult questions on the science of *measurement* than ever before. Most of this newfound energy and enthusiasm stemmed from Captain William Sampson, the chair of the department of physics and chemistry. Among other things, perhaps under the influence of Harvard astronomer Edward Pickering, he introduced new and more sophisticated course work as well as greater student involvement, even to the point of encouraging them to make suggestions for improving *the apparatus*. He noted with pride that "in almost every measurement, some portion of the apparatus has been improved by each student." In short, Sampson moved far ahead of his time by his extensive use of laboratory work in college science courses.

THE MICHELSON PERSONALITY

Michelson had his own way of existing in the world. He cared not a jot about areas of human knowledge outside of physics. In his own field, however, he kept up relentlessly. He could be bullheaded, easily forming strong and unshakable opinions. His reaction to the discovery of X rays by Roentgen was typical: "I confess I do not see . . . how any important scientific or even practical application can be made of it."[3] On still other occasions, he could be unimaginably quiet and mysterious to the point of inscrutability. Although not a social pariah (he had dated at Annapolis and eventually married Margaret W. Heminway, the niece of one of the academy staff whom he'd met while at Annapolis), Michelson's universe revolved around his work. What happened outside meant little, save for rare exceptions such as the sinking of the *Maine* by the Spanish in 1898, when Michelson spoke at the University of Chicago and demanded that the United States declare war on Spain. The battles between management, government, and labor were nothing compared to his battle to grasp the essential nature of the universe.

HEADING FOR EUROPE

In the fall of 1880, Michelson headed for Paris to study optics with the French virtuosi (a term stolen from music, but applied to the best French scientists). Hardly a surprise. European science was winning great international acclaim that year. Josef Breuer, later to be so influential to Freud, used hypnosis for the very first time that year to treat hysteria and Lord Kelvin was writing his great work "On the Size of Atoms." It was thought absolutely necessary then for an American science student to complete his studies in Europe to be competitive. Thus, it was common for American students to make this pilgrimage to study under masters such as Helmholtz, Thompson, and others.

But this requires some comment, as there has been much misunderstanding about the dependence of Americans on European science. For one thing, after the turn of the century, fewer and fewer Americans went abroad to study physics. Instead, more Europeans started coming here. One of the things that made this possible was the growth of summer schools in the United States, although the growing prestige of American science and scientists also acted as a lure. Ludwig Boltzmann, for example, visited Berkeley in 1905, while Rutherford lectured at Berkeley in 1908, and Max Born traveled to Chicago in 1912.

There is a difference of scholarly opinion here: most historians of science lean toward the view that American science was still greatly inferior to European. However, historian K. M. Olesko has argued that some have exaggerated the differences between German and American science and, in Michelson's case, that "the shaping of Michelson took place while he was an instructor at Annapolis and had been completed before he left for Germany as well as before he was hired by the US Naval Observatory in the fall of 1879 to work with Simon Newcomb."[4] I lean toward Olesko's view, which is to say that I believe that the United States, while perhaps still lagging behind European science, was starting to catch up rapidly. But the question is still open and resolution has to await further research.

MICHELSON MOVES ON

Soon, Michelson had the chance to take a post at the Case Institute of Technology. The Case school was the result of the munificence of Leonard Case, Jr., an unfathomably rich Cleveland area businessman and an enthusiastic amateur scientist. It too had its stars. One of these, Edward Morley, would attain immortality in his later work with Michelson. But Clark University wanted Michelson also. Clark had opened in 1887, with a mission primarily in research, rather than teaching. Its founder, Jonas Clark, acting with the advice of Woolcott Gibbs and Cornell's founder Andrew Dickson White, began the university only as a graduate program in science, specializing in physics, psychology, and biology, among other fields. However, Clark wanted more than good programs: he wanted prestige. At this point, only Johns Hopkins and Cornell carried such international prestige and Johns Hopkins psychologist G. Stanley Hall wanted to catch up as soon as possible. Indeed, Hall, its first president, was as eager as Clark to build the school's reputation.

Thus, acquiring Michelson was a priority. Case had appointed Michelson as professor of physics in 1881. Fortunately for Clark, Michelson had been growing increasingly dissatisfied with the caliber of students at Case and had been having severe problems with the administration over his handling of the departmental budget. Things soared beyond Michelson's tolerance level when a fire devastated the Case laboratory of his collaborator Edward Morley (which Michelson had already been using). Irritated by the slowness of the Case administration to rebuild, he accepted Clark's offer in May 1889.

By 1892, Michelson's reputation had grown enough, largely because of his research on theoretical questions related to the refraction and reflection of light in the 1890–91 academic years, for him to be invited to be on the faculty of the fledgling University of Chicago. Founded in 1881, its head was the brilliant and ambitious William Rainey Harper, the son of an Ohio storekeeper and former professor of Hebrew theology at Yale. He was as interested in good teaching as in good research. (Though Michelson was later exempt because he was a "head professor," he did lecture occasionally in freshman and sophomore courses as a guest teacher.)

Michelson, however, was at first hesitant. For one thing, the money was not terrific. As with most university presidents of the day, money had indeed been a problem throughout Harper's reign. And attracting "name" faculty required big money. To alleviate the financial pressure, he pursued, among others, John D. Rockefeller. In this way, he obtained hundreds of thousands of dollars for the new university. The first freshman class had over 200 graduate students and 1000 applications for faculty jobs.

A turning point came, however, when the distinguished biologist Charles Whitman, the chair of the biology department of Clark, decided that he would be unable to come to terms with Clark University. He disagreed sharply with its president's inclination to run each department over the heads of its chairs. Finally, Whitman decided to accept Harper's offer.

Soon, at a meeting at Whitman's home, Harper spoke to Michelson as well, convincing him (as well as several other disaffected faculty at Clark) to go to Chicago. When the dust cleared, Harper had "stolen" about half of the faculty at Clark and Michelson became the first "head professor" of physics at the University of Chicago.[5]

But problems continued to erupt between Michelson and Chicago. Because of his fame, the university had high hopes that Michelson would attract many graduate students eager to study at the master's knee. That is exactly what happened. But the problem was that the university had not carefully appraised Michelson's personality. His short temper and impatience were notorious. He disliked giving of his time, and he especially disliked giving it— and his hard-fought wisdom—to "partially educated" graduate students.[6]

Fortunately, there was Robert Millikan, of a different temperament than Michelson. Michelson soon brought Millikan there as a research assistant, placing the burden of reading graduate theses in Millikan's hands. Indeed, in the years 1892 to 1902, Michelson taught only three graduate courses—on spectral analysis, interferometry, and the velocity of light. As a result, by 1921, according to the historian of science J. L. Michel, only two of sixty-five graduate students had expressed interest in working under Michelson. Later, Millikan reminisced about Michelson's attitude (I here give Millikan's version, but it should be noted that this quote has often been reproduced in slightly varying forms):

> He had had a number of unfortunate experiences with the research problems which he had assigned to graduate students, and after a particularly bothersome bout with one in 1908 he said to me, "I don't want to turn over any more of my problems to graduate students. They always do one of two things. They either mess up my problem so that I cannot take it away from them and do it myself. Or else they get some real results after I have told them just how to do so and then begin to think the problem is all theirs instead of mainly mine. But the fact is that recognizing a good problem is the major part of research. So this is what I propose to do: I'm going to hire an assistant who will do just as I tell him, and will not expect any credit for himself or make any demands upon me other than that I keep his weekly pay check going. If you will take over all the assigning of thesis subjects I'll be your debtor forever.[7]

From the time Millikan took over, Michelson became even more of a shadow figure in the physics department. He rarely attended meetings and generally spoke only with his close friends. Like many scholars, Michelson had minimal patience for administrative and routine paperwork. Any spare time he had he played pool at the Quadrangle Club, or practiced his scales as a none-too-skillful violinist.

Still, one can overstate all of this. He did, for example, have a *few* graduate students who were successful enough. In 1890, Louis Austin and T. Proctor Hall worked on applications of interferometry to measurement problems. Hall was probably the best of Michelson's students, even though he finished his doctorate at Clark University. He was noteworthy in that he had developed, under Michelson, new methods for measuring surface tensions to a far higher accuracy than ever before.

But even if Michelson did wall himself off from graduate students and even other faculty members to some extent, he did eagerly engage in correspondence with other distinguished scientists. The "scientific correspondence" in the Gibbs collection at Yale University shows, for example, that he had extensive discussions (six letters from Michelson) with J. Willard Gibbs, certainly America's first great theoretical scientist, on the nature of light and light waves. Indeed, the first letter in the series shows Michelson discussing the Michelson–Morley experiment, though none of Gibbs's replies have survived (assuming he did reply to Michelson).

The first letter, dated December 15, 1884, says,

> My dear professor Gibbs . . . granting that the effect of the atmosphere may be neglected, and supposing that the earth is moving relatively to the ether . . . would there be a difference of about one hundred-millionth in the time required for light to return to its starting point . . . when the direction is at right angles? . . .[8]

THE QUEST FOR THE SPEED OF LIGHT

Since Mesopotamian and Egyptian times, light has always been some kind of "ultimate" metaphysical symbol, often for the awesome powers of the cosmos. So it had been for the followers of the cult of Aton in the last important period of ancient Egypt's history, New Kingdom (1560–1087 B.C.), and so it was for Michelson. From the moment he began his research, it was evident that Michelson was not in search of ordinary or minuscule achievements. Measuring the speed of light is not a feat for the timid, nor for those of small vision.

Since the belief for centuries was the quite natural and intuitively plausible idea that light traveled "instantaneously," science had to wait until the 17th century for anyone to dream that things might be otherwise. However, Danish astronomer Ole Roemer had noticed that one of Jupiter's moons cast a shadow that appeared intermittently on the surface of the planet, exactly 16 minutes and 36 seconds earlier than at other times. He judged that this might be so because at one time of the year the Earth was closer to Jupiter than it was six months later. He then theorized that the time light took to traverse the Earth's orbit might account for the time difference. And when he divided this distance by the 16 minutes and 36 seconds, he arrived at the first close, but inaccurate measurement of light—180,000 miles per second. In fact, it was to be over 180 years before scientists would be able to measure the speed of light with even approximate accuracy, by measuring the time it takes to pass between two points on Earth (just as one would calculate the velocity of a race car, by measuring the time required for it to pass between two points). In a similar way, French scientist Armand Fizeau, relying on a known procedure, also measured the speed of light. He did it by sending bursts of light to a mirror and measuring the time the reflected light took to return. This was the first step toward the development of the interferometer, a device that splits a light beam in two in order to measure its speed. Michelson first conceived of it in 1881.

As it happens, Simon Newcomb, the well-known American astronomer, had, with congressional aid, been tackling Fizeau's old problem of measuring the speed of light. But in 1877, Michelson worked out a new procedure that would measure light with the greatest simplicity. He published his findings in the *American Journal of Science* in May 1878. This was the same journal that in 1887, under the editorship of James and Edward Dana, would publish Michelson and Morley's findings on both the ether and the new meter standard. Instead of working with a distance of only 23+ meters, Michelson extended the distance to 700 meters. He further improved the measuring apparatus so that he managed to increase the deflection of the image to 13.3 centimeters.

In a letter from Maxwell to David Todd of the Nautical Almanac office, Maxwell had been inquiring about information on the orbits of Jupiter and its satellites which could show the motion of the Earth through the ether. In the letter, Maxwell claimed that it was impossible to measure the speed of light sufficiently accurately to notice any movement of the Earth through the ether.

Michelson flatly disagreed and the challenge merely intensified his interest. Since Maxwell had claimed that light would turn out to have a speed of 300,000 kilometers per second, or

186,000 miles per second, it seemed to many to be a near-miracle when Michelson obtained the result of roughly 299,895 plus or minus 30. In effect, he had proven Maxwell to be very nearly a prophet by confirming his prediction with an accuracy of 1 in 10,000. (Michelson's father-in-law deserves some part of the credit: he had contributed $2000 for this work. Michelson used this money to build a house specifically designed to conduct his speed of light experiments, one with equipment adequate to control the temperature.)

So it was that, in April 1878, Michelson published these results, based on an earlier talk to the AAAS. It was not to be the last time Michelson would attack this problem. Even by the age of 74, he would try a system involving sending a shaft of light from Mt. Wilson over 20 miles to Mt. San Antonio and back again. At that time, he constructed his rotating mirror with the highest accuracy that the technology of the 1920s allowed. But although he verified his earliest measurement of light, he did not substantially improve on it. Nonetheless, he spent a major part of the remainder of his life trying to do so.

THE ETHER

For a number of years, Michelson had been thinking about the idea that all light travels in that mysterious medium called the "ether," referred to in the above letter to Todd. Scientists universally assumed that the wave motions of light took place in, and were transmitted by, this substance. Most scientists believed it was also immovable, while pervading all of space throughout the universe. The ether was a concept whose roots dated back at least as far as the Chinese Taoist philosophers of the 6th century B.C. The founder of Taoism, Lao-tzu, had postulated the "flowing Tao," an impalpable force that also pervaded all of space and all persons and which explained all happenings in the universe. Like the Tao, the "ether" was impalpable and omnipresent and the basis for many explanations of change of any kind in the physical world.

Telegraphy researcher Oliver Lodge offered a reasonably accurate conception of the ether when he described it as:

> one continuous substance filling all space; which can vibrate light; which can be sheared into positive and negative electricity, which in whirls constitutes matter, and which transmits by continuity and not by impact every action and reaction of which matter is capable.[9]

Indeed, Newton's colleague and occasional adversary, Christiaan Huygens, had published the idea as early as 1690, during Newton's era, though Maxwell had subsequently refined and developed the concept. The ether acquired even more adherents with Maxwell's successes in unifying electricity and magnetism (the first step, arguably, to what physicists today call a "grand unified field theory"), believing that the ether was the *locus primus* of the electromagnetic force.

Michelson had been contemplating the sort of experiment needed to test this concept. Soon he had a good idea—an idea for the type of equipment that, as has often been claimed, would lead inevitably to the refutation of the ether theory and, ultimately, to Einstein's theory of relativity. (Many historians now doubt this latter claim.)

To understand the significance of this phase of his work, it is essential to understand the state of the world in the 1870s. A growing mythology at the time held that the future of physics would be merely that of finding small improvements in gadgetry and new practical applications of the laws of physics. Many scientists, including some of the most celebrated, such as Lord Kelvin, believed that all of the laws of science had already been discovered. Michelson also held this view—for the moment. Indeed, the intellectual climate did seem to warrant this eminently "practical" approach. Edison was nearing the height of his powers as was George Westinghouse. Edison would perfect his system of "multiple" telegraphy (sending more than one message simultaneously over a single wire) while George Westinghouse, at the same time, perfected his railroad air brake.

But Michelson soon discovered that if scientists had answered the key questions, the questions themselves were wrong. Some of the most widely accepted concepts in physics would soon collapse in ruins. Among the debris would be found the stationary "ether."

THE INTERFEROMETER

The next step in Michelson's efforts to settle the ether question once and for all involved the "interferometer." Michelson's idea made use of a common phenomenon in nature: it was the same phenomenon that told why colors of the spectrum could be seen on a film of oil coating the surface of water. Although much of the light bombarding the surface is reflected off of the oil film, some light does enter the surface and is reflected only when it hits the lower surface. At some angles, the two light reflections interfere with one another—a phenomenon common to wave motion of any kind. In the course of this interference, some colors cancel each other and the oil's surface appears black. But where the colors reinforce one another, colors appear on the oil's surface. (That's why you often see "rainbows" on sidewalk puddles of motor oil.) The interferometer worked on just this principle. It divided a light beam in two, just as the two surfaces of oil film will split sunlight.

With the interferometer, light beams could be split and sent in two separate directions, rejoining again on their return. Thus, one beam would travel parallel to the Earth's rotation, while the other went perpendicular. When they returned, the two beams would reassemble, producing clearly visible "fringe" patterns, indicating that the two beams were interfering with one another.

The key to the *determination* of the speed of light consisted in the fact that the two light beams traveled at right angles to one another before coming together again. The slightest difference in the paths of the two beams would correspond, in the above oil analogy, to the difference between the light beam reflected from the upper surface of the oil and from the lower surface. In both cases, the experimenter would be able to see chromatic patterns— alternating dark and bright bands of light. Thus, if one beam headed in the direction of the Earth's path in space, which meant traveling in the ether, while the other beam headed in space at right angles, there should have been *some* detectable difference between the two paths.

Knowing the frequency of the light beams, Michelson would have been able to calculate the vanishingly tiny difference between the paths of the two beams. (An analogy here would be a situation where two strong racers headed into a race with a powerful wind. One racer would be instructed to run across the wind and back, while the other would be instructed to run one mile downwind and back. Clearly the runner heading across the wind would be dragged to the side a little on both crossings and would thus end up farther down from the point he had started. The second runner would travel with the wind rapidly and could return only with great difficulty. So, the runner heading upwind and down, would take more time than the one heading across the wind. Starting with the time taken by both runners, the wind speed can be calculated precisely.)

Although that was the expected result, it didn't happen that way. And this finding forever changed all future science. At this epochal moment, Michelson noted that none of the expected changes in the interference patterns showed up. Had there been an "ether," they would have. Since they didn't, there was no "ether." It was a staggering blow to some ideas about the concept of an ether, a concept that undergirded virtually every scientific conception of the day. However, Michelson did not question the *existence* of the ether in the 1881 paper—only Fresnel's idea that the ether remains at rest. However, some interpreted the experiment as showing that the ether does not exist at all. Michelson discussed all of this in his classic paper that appeared in the *American Journal of Science* in 1881. (This journal, published out of New Haven,

Connecticut, was virtually the only important journal for physicists in the United States. Among many of its features, it presented article summaries from important foreign journals, such as *Annalen der Physik*, *Acta Crystallographica*, and others. The great *Physical Review* would not appear for several more years.) In this essay he said merely that "the hypothesis of a stationary ether is thus shown to be incorrect."[10]

It was scarcely a mystery that scientists immediately assaulted this drastic conclusion. Physicist Henry Rowland dismissed it, believing it inconceivable that light could travel across empty space without some kind of medium in which to travel.[11] British doubters wanted the experiment run again. For a while it seemed like everyone wanted it run again.

THE LORENTZ–FITZGERALD
CONTRACTION HYPOTHESIS

Finally, under pressure, Michelson agreed. Perhaps it was the scientific atmosphere: progress in the understanding of the nature of light and electromagnetism generally was coming rapidly. That year, 1887, for example, was the year that Heinrich Hertz and Oliver Lodge showed that radio and light waves are merely different regions of the electromagnetic spectrum. So, with the cooperation of another Case faculty member, Edward Morley, Michelson repeated the ether test. Again he had a null result.

However, other new and powerful ideas emerged. Irish physicist Edward Fitzgerald and Dutch physicist H. A. Lorentz had an idea for "rescuing" the ether. They suggested a concept later to become known as the Lorentz–Fitzgerald contraction. Their idea required the bizarre notion that objects, such as one arm of the interferometer, in motion against the ether could *shrink* in the direction in which the arm was moving, thus accounting for the apparent null result of Michelson's experiment. As Lorentz put it, "the line joining two material particles shifting together through the ether"[12] contracts to explain the null result of the Michelson experiment. (It should be mentioned that during Lorentz's time, physicists considered Lorentz's move arbitrary and "ad hoc"—merely a desperate device to keep old ways of thinking alive. Now, however, many historians of science are not so sure that Lorentz's idea was so ad hoc, and believe that it was a quite logical requirement of both the theoretical and experimental results of science at the time.)

The degree of contraction was, in turn, a function of the object's approximation to the speed of light. The closer the speed of the object to that of light, the greater the contraction. Thus, at everyday velocities, the degree or amount of contraction would be negligible. But at, say, one-third the velocity of light, the shrinkage would be substantial—as much as eleven percent.

To many, accepting such a bizarre notion was hardly more plausible intuitively than Michelson's dismissal of the ether. Thus, many, although by no means most, scientists rejected the "contraction" hypothesis out of hand. The German physicist Kaufman, however, in 1901 demonstrated that the Lorentz idea did seem to apply, at least to subatomic particles. He found that the electrons emitted in radioactivity increased their mass as their velocity increased. Further, the pattern seemed to follow the Lorentz theory—the greater the velocity of emitted electrons, the greater the mass. Such discoveries portended the beginning of the end for the invincibility of the old classical physics of Isaac Newton. And, with the great Max Planck publishing his quantum theory of radiation that same year, the "new" physics of quantum mechanics and relativity began to appear.

EINSTEIN AND MICHELSON

Although Lorentz's defense of the ether proved to be incorrect, part of his conception would be incorporated, in 1905, into a theory that would shake the world—the theory of special relativity.

Here I think a caution and brief digression is in order. Though I tend to believe this latter claim, it has to be noted that many historians of science deny it. Many think this alleged link between Michelson's work and the theory of relativity to be at best a huge exaggeration. On the other hand, historians Barbara Haubold, Hans Haubold, and Lewis Pyenson in their relatively recent article "Michelson's First Ether Drift Experiment in Berlin and Potsdam," argue for this link.[13] Jed Buchwald, in his essay "The Michelson Experiment in the Light of Electromagnetic Theory Before 1900," doubts that the experiment changed many minds. He claims that ". . . most physicists, including Henry Rowland, in the period immediately following Michelson's work, did not abandon the ether theory, nor even downplay it."[14]

Also, historian of science L. S. Swenson, in his article "Michelson–Morley, Einstein and Interferometry," claims that history has misrepresented the impact of their famous work:

> What is surprising, to me at least, is how flimsy was the actual character of the 1887 Michelson–Morley experiment compared with its widespread and formidable reputation. Already . . . many of those who wrote and spoke about the significance of their experiment after them [Michelson, . . . Einstein and others] were not [honest in checking their sources . . .].[15]

(Swenson is not implying foul play here, merely carelessness on the part of subsequent scholars. Part of the complications were due to the fact that another scientist, Dayton C. Miller, who had a Princeton Ph.D. in astronomy and who had collaborated with Morley three times as long as Michelson had, actually appeared to have obtained some evidence that the ether existed. Later, however, according to Swenson, the scientific community agreed that Miller's positive result was due to design flaws in his experiment, including Miller's failure to monitor temperature properly.)

Granting all of this, Einstein's own words do seem to make it clear that at least he took Michelson and Morley's work very seriously. In his article, "How I Created the Theory of Relativity," he said:

> While I was thinking of this problem [of the motion of the Earth through the ether] in my student years, I came to know the strange

result of Michelson's experiment. Soon, I came to the conclusion that our idea about the motion of the Earth with respect to the ether is incorrect, if we admit Michelson's null result. This was the first path which led me to the special theory of relativity.[16]

I do not think an absolutely decisive case has been made either way. In any case, on the assumption that the work was important for relativity, Einstein would later show that both Michelson's and Lorentz's concepts could be reconciled by using newer and even more dramatic ideas. Ironically, although the Lorentz–Fitzgerald hypothesis was indeed successfully woven into the fabric of Einstein's special theory of relativity, Lorentz himself never accepted some of Einstein's ideas, such as the "relativity of simultaneity," the idea that it was meaningless to speak of two events as "simultaneous" except from the perspective of a specific frame of reference. He always claimed, against Einstein, that there was no difficulty in distinguishing simultaneous and nonsimultaneous events, regardless of the reference frame.

The foundations of special relativity had appeared: it remained for Einstein to complete the edifice. According to Einstein's famous 1905 paper, all points in the universe are equal for purposes of a starting point for measurement. In other words, there are no "absolute" positions in space, no absolute "up," "down," and so forth, as Newton had thought. (This is somewhat analogous to a popular view of ethics that holds that there are no "absolute" ethical truths—one person's ethics is as good as anyone else's.) The laws of physics are therefore identical in all systems possessing uniform motion with respect to each other. He also showed that the velocity of light is independent of the velocity of the source of light and the observer. Thus, an observer confined to one system cannot detect the motion of that system, no matter what experiment he or she performs. Put differently, no matter who measures the speed of light, nor under what conditions it is measured, you always get the same answer. Example: Newtonian "commonsense" physics held that if, say, you were traveling in a jet at 700 mph and shining a light beam toward the tail of the plane, the speed of the light beam would be the speed of light plus the speed of the airplane. Einstein, by contrast, showed that the speed of light would remain the same, whether seen from a moving plane, the ground, or anywhere else.

One can therefore measure motion only relative to an observer performing the measurement. Einstein's special theory of relativity in this way demonstrated that the "contraction" hypothesis of Lorentz and Fitzgerald was a correct description of reality. However, Michelson's experiment, at least according to some possible interpretations of it, offered some evidence that the contraction was unrelated to a nonexistent "ether." As a corollary to all of this, Einstein, influenced by Fitzgerald and Lorentz, also showed that the mass of a physical object would increase as its velocity increases.

Ten years later, Einstein extended all of this, producing the general theory of relativity, which was subsequently supported by astronomical observations. (Though not so easily as is often supposed. In the Millikan Library at Caltech there is a Science Service release of August 16, 1933 titled "Star Photographs Leave Einstein Effect in Doubt," showing that astronomers believed that "there exists a definite discrepancy between the Einstein theory and the results of total eclipse observations, since the plates obtained by the . . . camera yielded a deviation about 25 percent higher than the [expected] value." Later and more refined astronomical photographic methods finally yielded data in better agreement with Einstein's general theory.)

I would like to offer a comment on Edward Morley. Although the famous ether experiment is commonly referred to as the "Michelson–Morley" experiment, Michelson usually gets most of the attention. Perhaps this is fair enough, considering the sum of Michelson's contributions. Yet Morley was no bystander either. According to historian of science D. H. Stapleton, "any account of the Michelson–Morley experiment indicates that Morley's role was considerably more important than most writers had suggested."[17]

THE NEW METER STANDARD

The interferometer had many other uses. Reworked to coordinate with telescopic observations, its use led to a number of astronomical measurements, such as the calculation of the size of the moons of Jupiter at the Lick Observatory in 1891.

And with at least one application of interferometry, finding a new meter standard, a darker side of Michelson's career emerges: he may have unfairly pushed Morley out of the picture. The International Bureau of Weights and Measures had been established in Paris in May 1875, the year before the patenting of the telephone. The scientists at the bureau had been following Michelson's work for some time, hoping to substitute a standard based on the wavelength of light emitted from a variety of hot substances, in order to make the concept of a meter uniform worldwide. In a paper published jointly with Morley, they had, in fact, already suggested the possibility of basing a standard on something other than the old platinum/iridium meter bar.

Sadly, it was at about this time that Michelson's relationship with his old friend Morley soured permanently. While Morley wanted to persist along these lines, Michelson wrote him in December 1889 saying only, "I am sorry that I have to work alone at the wave lengths." It was a distant cry from a mere two years earlier when Morley wrote optimistically to his father, in April of that year: "Michelson and I have begun a new experiment. It is to see if light travels with the same velocity in all directions. . . ." Indeed, Michelson and Morley collaborated for barely five years, though they accomplished a lot during that time. Why this change of attitude on Michelson's part? It is not completely clear. However, friction between the intensely religious Morley (who had been a minister at one time) and the agnostic Michelson doubtless contributed to the breakup.

In any case, Michelson forged ahead alone. He first tried such substances as mercury in his quest for a new standard, but he decided that the spectral lines emitted by mercury on heating were far too complicated to make accuracy feasible. Cadmium, on the other hand, emitted simple and narrow lines, making it easy to work with. The work ended in the spring of 1893, the year Henry Ford built his first automobile. The world had a new standard meter for measurements of length, set at 1553, 163.6 wavelengths of the red cadmium wave line. (A further critically important consequence of this, not fully appreciated at the time, was that Michelson realized the great complexity of many light spectral lines, or what is now called "hyperfine structure." This concept was to prove invaluable in nuclear physics later on by such giants as Fermi, Urey, Bethe, Soddy, and others.)

MICHELSON'S WORK ON THE ELASTICITY OF THE EARTH

In 1919, Michelson undertook perhaps his greatest challenge to date: measuring the rigidity of the Earth to explain tide formations. The great unanswered question was whether the depths of the Earth were solid or liquid. Other scientists interested in the problem were Thomas Chamberlin, Fred Pearson, Tom O'Connell, and Henry Gale at the University of Chicago.

Once again Michelson's engineering prowess came through. He engineered two pipes, 300 meters and 600 meters, with a radius of 3 inches. He then buried them 10 feet underground at right angles to each other. Next he half-filled them with water. Then he set up still another gadget he'd designed himself, an optical-fringe device (based loosely on the idea of the interferometer). With it, he was able to measure precisely the shifts in the water levels in the pipes, as minitides were produced in the pipes by the gravitational pull of the moon and the sun.

The theoretical idea was as follows. If the Earth were, say, solid rock and thus completely rigid, he would be able to detect small tides in the pipes. On the other hand, if the Earth had a relatively low index of rigidity (which would be the result of its being partially liquid in composition), he would observe no change in water levels in the pipes. In fact, he did observe such tides, minuscule though they were, indicating to him that the Earth was more liquid than solid—a theory later confirmed by geologists. (In effect, this was a visual analogue of the famous Foucault pendulum, an ordinary pendulum that shows the Earth's rotation. The original consisted of a 28-pound lead ball suspended on a metal wire 67 meters long.)

OLD AGE AND LIGHT MEASUREMENTS IN A VACUUM

With the coming of WWI, at age 65 and with much of his great work behind him, Michelson rejoined the Navy and was given the rank of lieutenant commander. His assignment was to develop hardware that would be militarily useful. In fact, he devised a range finder that would become widely used by all four branches of the armed forces for many years.

But without question, the major undertaking of his later years was his series of measurements of the speed of light in a vacuum. Backed in his efforts by the Rockefeller Foundation, the Mt. Wilson Observatory, the University of Chicago, the Carnegie Corporation and the Coast Guard, Michelson embarked on this bold new venture in 1928. In that year he first measured the velocity of light in a vacuum. For this monumental work, he was offered a post at Caltech. As he commented later, "Hale said I could have Mount Wilson and Caltech. . . . The temptation was too great. So I came." The point of this experiment was to measure light without the encumbrances that had plagued him in the past—smoke, smog, fog, clouds, etc.

But his biological clock was inhibiting his ambitions. By 1931, he was nearing 80 and his health was deteriorating rapidly. He knew that death was near. As a result, his colleague Pearson had to do much of the actual work, including adjusting the apparatus, making the actual measurements, and correlating the data. Michelson, before drifting into a coma, carried on bravely, saying only, "My health continues to improve." He died two days later.

But beginning in 1930, Pearson, under Michelson's guidance, had managed to make literally thousands of observations. The final result: Michelson and Pearson settled on the figure of 299,774 kilometers per second as the speed of light in a vacuum, or just over 186,000 miles per second. The title of his final paper on the subject, "On a Method of Measuring the Velocity of Light," turned out to be the same as his first, published at Annapolis in 1878. (This was probably unintentional, although no one seems to know for sure.)

Michelson held the post of president of the American Physical Society from 1901 to 1903, and was president of the American Association for the Advancement of Science for four years. The number of lifetime awards given to him was staggering—the Copley medal of the Royal Society, the Draper medal, and the Nobel Prize in 1907, as well as innumerable honorary degrees from universities all over the world.

He occupied a pivotal moment in history, extending from the closing days of the 19th century to the difficult and turbulent war years at the early part of the 20th century. By 1932, Hitler, at the height of a nationwide banking collapse in Germany, was already powerful enough to refuse Hindenburg's offer to become vice-chancellor. Gandhi was at the height of his influence in a turbulent and poverty-stricken India.

In 1931, Edison died. Again, the timing was symbolic. For 1931 and 1932 were great and, indeed, almost miraculous years in physics. Quantum mechanics was only seven years old and developing rapidly. In 1931, E. O. Lawrence invented the cyclotron or "atom smasher" and by 1932 Carl Anderson had discovered the positron or "positively charged electron" (electrons normally bear a negative charge). And there was much more: James Chadwick found the neutron and Irving Langmuir won the Nobel Prize in chemistry for his work on the chemistry of surfaces. Amelia Earhart became the first woman to fly solo across the Atlantic.

Michelson saw too the evolution of American scientists from a group of relatively lackluster apprentices compared to the greats of Europe, to a group of men as ingenious and innovative as any of the titans of the European universities. He witnessed the gradual evolution of American science from the backyard, amateur enterprise it was throughout the 18th and 19th centuries, to a full-fledged profession with journals, professional societies, peer-review systems, and government funding. He witnessed the earliest stirrings of quantum mechanics, one of the great revolutions in the history of the physical sciences and, indeed, of humankind.

When he died, his colleague, Robert Millikan, wrote a brief biography and testimonial to his greatness in *Scientific Monthly*. Dying when he did, Michelson was spared the horrors of the atomic age, horrors that would profoundly affect many of his colleagues, including Pauling, Fermi, Teller, and Feynman, among many others.

CHAPTER 5

R. A. Millikan and the Maturity of American Science

In September 1886 a teacher wrote on a Maquoketa High School report card: "Robert Millikan gained an excellent standing for good deportment, faithful application to duty and exceptionally high grade of scholarship."[1] As time soon told, this student did not disappoint. Later in life, he would, for the first time in history, accurately measure the charge on an electron, verify Einstein's theoretical ideas about the nature of light and contribute enormously to our understanding of those mysterious visitors from space—cosmic rays. Beyond all of this, he would midwife the growth of the California Institute of Technology from its beginnings as the small and undistinguished "Throop" institute to the world-class institution it is today.

Robert Millikan, born in 1868 in Illinois, was the son of a minister and was raised in the tiny village of Maquoketa, Iowa. His boyhood was normal enough: he and his brothers and sisters regularly arose at three in the morning to milk the cows, play baseball, swim, and play at gymnastics on their homemade equipment.

As in many midwest schools during this era, science teachers were scarce and often poorly trained. Millikan's high school was no exception. The principal of the school taught the one course in the field, such as it was. He believed in divining rods and had little faith in Isaac Newton or any of the currently accepted theories in physics.

Nevertheless, Millikan did well enough to enter Oberlin College, in Oberlin, Ohio, at the age of eighteen. True, acceptance had been easier because his granduncle had helped found the institution and it was, in any case, the alma mater of both of his parents. Oddly, a professor of Greek, John Fisher Peck in the Department of Preparatory Studies, was one of the earliest important influences on him, as he was the first to encourage Millikan to study physics. As Millikan said later in life, "I owe a greater debt of gratitude to him than to any one outside of my own immediate family, for the direction my life has taken."

Though Millikan's own interest in physics was still marginal, other faculty at Oberlin were beginning to appreciate his vast talent in science and offered him encouragement. Eventually, he was able to expand his scientific background with courses in physics, math, German, and geology. Later in life he recalled his feelings of insecurity, "I was terribly blue because the other men whom I admired got their work so much more easily than I got [understood] mine." Apparently, Millikan was a comparatively slow learner in school, at least in the earlier grades.

ON TO GRADUATE SCHOOL

Notwithstanding his spotty background, Columbia was impressed. His transcript showed extremely high grades, even if Millikan had to work very hard to get them. Columbia offered him a fellowship of $700, which was impressive enough that he accepted it. At this point in his life, Millikan had already started becoming seriously involved with physics. Since Columbia lagged well behind institutions such as Cornell and Johns Hopkins in physics, he could enter it with at least some degree of confidence.

Still, Michael Pupin was there. During World War I, Pupin had been a member of the National Academy of Sciences. He was also an enormously successful inventor, whose work had brought AT&T to the mainstream of American industry. Among his other accomplishments, Pupin was renowned for his virtuosity in experimentation as well as his vast knowledge of magnetism. He had graduated from Cambridge and studied with German physicist Hermann von Helmholtz, who first suggested that there was a smallest unit of electricity, and German physicist Gustav Kirchhoff, a pioneer in the study of the relationship between electricity and magnetism. It was hardly surprising that Millikan very much admired Pupin.

Other physicists at Columbia were Robert Woodward and Ogden Rood. Though these were lesser minds, they too inspired Millikan. With their encouragement, his own interest matured and he ultimately decided that physics would be his lifework.

OFF TO EUROPE

Pupin wanted Millikan to study with Helmholtz, so Millikan headed for Europe with $300 lent to him by Pupin (albeit at 7% interest). Even though American science was growing more professional, Europe was still the only place to go for high-level theory. European physicists still considered America to be an "experimentalist" environment. This does not necessarily mean that American scientists sought "practical" applications for theory, but merely that they believed experimentation rather than theory-construction was the most important part of science. And since the world's elite in physics were in Europe and did not hold that view of experimentation, they considered an American education less prestigious than a European. (Later, America would catch up in theoretical science.) Most American doctoral dissertations involved some kind of Michelson-type precise measurements, whereas European dissertations tended to be more theoretical. At

Johns Hopkins, for example, the late 19th century dissertations usually had something to do with Rowland's experimental ideas in spectroscopy.

In preparation, Millikan began studying German. Once there, he attended lectures in Paris by French mathematician Jules Poincaré and lectures in Germany by German physicist Max Planck, who first suggested the idea of a quantum, or the smallest unit of a physical property, such as energy or momentum. He was in Germany for example in 1896, when Roentgen discovered X rays. In fact, Millikan was among those who saw the very first X-ray photos ever taken. (Soon, using this new technology, George Eastman would revolutionize photography and, in particular, astronomical photography.) But Millikan didn't tarry, hurrying along to Göttingen to study under Walther Nernst at his institute for electrochemical research.

The most important aspect of his European pilgrimage was that, at the time, the important controversies over the nature of cathode rays were taking place. Eventually, the discovery of the electron by the English physicist of the Cavendish Laboratory at Cambridge, J. J. Thomson, resolved all of these controversies. This first drew Millikan to the study of the electron and, ultimately, to his famous measurement of the charge on an electron.

THE DISCOVERY OF THE ELECTRON

In 1897, Thomson published his classical treatment of *cathode rays*, the name given to a flow of electrons in a vacuum tube, where he gave a complete synopsis of all previous work in this field, even suggesting some new approaches. Perhaps the most important—and most dramatic—suggestion was his idea that cathode rays were simply particles bearing a negative charge, whose mass was almost incalculably smaller than the nucleus of any atom. Thomson dubbed this new particle the "electron." Soon he shocked many by asserting confidently that a part of the Becquerel rays (this is really just the ordinary radioactivity seen in radioactive substances such as uranium or radium) were also electrons.

THE PHOTOELECTRIC EFFECT

At the time, because it was so counterintuitive, most scientists recoiled at the unfathomable idea that all matter was ultimately composed of discrete atoms, merely because the idea conflicted with commonsense observations that matter is continuous. Thomson's even more outlandish suggestions compounded their dismay even further. Predictably, virtually no scientists liked these ideas. They felt scarcely better about Thomson's idea that the new "electron" could explain the photoelectric effect, the phenomenon whereby the surface of a metal ejects electrons when one shines ordinary light on it. Thomson had fearlessly urged that this peculiar happening was merely an ejection of electrons from metal surfaces, caused by bombarding the surfaces with ultraviolet light.

This field is where Millikan would soon create his reputation. With an elegant and bewitching experiment, he would finally prove the atomic theory of matter. The main obstacle to progress was that physicists—even if they might allow that small particles carried electricity—shrunk in mortal terror from the even more radical idea that such particles had both a well-defined mass as well as an electric charge. Only very few believed that electricity was composed of the electrons Thomson discussed.

ON TO CHICAGO

It was during this formative period that Millikan met Michelson at the new University of Chicago. Michelson was, of course, already well known by this time. He'd achieved recognition for the "Michelson–Morley" experiments from 1878 to 1890 which allegedly disproved the existence of that mysterious substance in which light waves were thought to travel—the "ether." Michelson had left Clark University in Massachusetts in 1892 to accept a post as director of the Ryerson Laboratory in Chicago.

Prophetically, Michelson contacted Millikan in the summer of 1896, while Millikan was still in Europe, offering him a position in his laboratory at the University of Chicago. Millikan was barely 28, but in his *Autobiography* he reflected on the University of Chicago: "The university was a new institution, strategically located in both time and place to exert a profound influence not only upon the evolution of the American University, but even more important, upon the kind of role which prospective universal secondary education was likely to play in American life."[2] Above all else, Millikan had to finish his doctorate and he went to Chicago for assistance from Michelson. Shortly thereafter, he enrolled in a number of Michelson's courses, including his course on the theoretical and historical aspects of physics, as well as Michelson's research course, reserved only for the best graduate students.

Michelson really wanted Millikan in his laboratory, having already seen his abilities firsthand. By the summer of 1899, Michelson allowed Millikan to take over the teaching of the junior college physics course. These lectures became the basis for his work, *Mechanics, Molecular Physics, and Heat*, published in 1902.

However, the sanction by Michelson proved to be both a blessing and a curse, for the cantankerous Michelson dumped onto Millikan the painful chore of dealing with graduate students, which Michelson himself could never stomach. Not surprisingly, the Ryerson labs under Michelson had a singularly undistinguished record for producing Ph.D.s. However, with Millikan to direct their work, there were a few. Henry Gale, for example, received a Ph.D. in 1899 for a thesis titled "The Relation between the Index of Refraction and the Density of Air." Later, Millikan and Gale would author an introductory text titled *A First Course in Physics*, published in 1906. It flourished for many years. (In fact, Linus Pauling's *high school* physics course used this book. In a 1964 interview, Pauling said, ". . . in high school I had one semester of physics, using a textbook by Millikan and Gale . . . and I liked this course."[3])

MEASURING THE CHARGE ON THE ELECTRON

In April 1902, Millikan married Greta Blanchard. Following a nearly yearlong honeymoon and study tour of Europe, he began an intensive period of study of the latest scientific advances, spanning the fields of cosmology, X rays, and the work of the Curies on radiation.

As the first decade of the 20th century was fading, Millikan began thinking seriously about the charge on the electron. So much depended on it and there had been no accurate measure taken of it as yet. In Millikan's own words, ". . . in October 1908 . . . I finished my first dozen years at Chicago and . . . was just getting into my most promising work on the absolute measurement of the electronic charge. . . ."[4] Millikan was hardly the first to attempt such a measurement. J. J. Thomson himself had tried in 1895. But despite many years of dedicated labor, he and his assistant Wilson had to confess that they were getting wildly different results with each attempt.

According to Thomson's "cloud method" as he called it, one could measure the mass of a body by measuring the gravitational force on a balance pan, a standard laboratory instrument for comparing masses of objects. As part of Thomson's method, physicist C. T. R. Wilson of the Cavendish Laboratory in England realized that clouds of water droplets could appear if he allowed water vapor to condense on ions (atoms or molecules bearing an electrical charge). First, scientists knew it was possible to calculate the mass of a particle by measuring its downward, gravitational force on a balance pan. Now, if a particle were electrically charged as Thomson claimed, and an electrical force pointing upwards could be put into equilibrium with the gravitational force, the particle should be exactly balanced. In that situation, it would be possible to calculate the electrical charge. And, if a single electron caused the charge, that charge would be the long-hunted charge of the electron.

Like any good scientist, Millikan first researched and reattempted Thomson's Cambridge method, since it was at least possible that some minor error was causing the varying results and that no radically new approach was necessary.

In fact, there was nothing wrong with Thomson's technique or idea. One of the stumbling blocks in all of this was that so far, no scientist had yet been able to design an apparatus adequate or sensitive enough to deal with individual particles. What they were doing was merely observing a cloud of charged drops of water and taking an *average* measure of the electrical charge of the aggregate of particles.

MILLIKAN'S APPROACH

Millikan had another supposition. He realized that the same apparatus could be used, if only he could adjust it so that it was possible to focus all of the measurements on a *single* droplet. With his assistant Louis Begeman, Millikan began reworking Thomson's approach. In one of the most significant early steps, he used radium, rather than X rays, to produce an ionized cloud, since the intensity of the X-ray beam was too difficult to control with the available technology of the day. In his words:

> To take the first steps in the . . . improvements, in 1906 I built a ten-thousand-volt small storage battery—at that time quite an undertaking—that would produce a field strong enough to hold the upper surface of . . . [the] cloud suspended 'like Mohammed's coffin.' . . . whenever the cloud was thus dispersed by my powerful field, a few individual droplets would remain in view.[5]

His intuition had guided him infallibly. By ridding himself of the other water droplets, he had thereby eliminated the droplets that had been throwing off the other measurements. And by increasing the power of the electrical charge, only the particles possessing the ideal balance of electrical charge and gravitational force remained. In short, he could now study the *relevant* particles individually and directly.

It has to be acknowledged that Millikan made the progress he did because of his efficient use of graduate students, at least during the oil-drop research. (It is called the "oil-drop" research because in later experiments, to achieve greater accuracy, Millikan substituted oil in place of the water droplets. He did this because oil evaporated far more slowly.) Some others, like Albert Hennings and Wilmer Souder, did their dissertations on work related to the photoelectric effect.

By September 1909, the measurements that would eventually capture the Nobel Prize for him in 1923 were near completion. Millikan felt ready to present his findings at the upcoming symposium of the British Association for the Advancement of Science at Winnipeg. Though he realized that the work was far from being either complete or definitive, he also knew that he was in the process of making history. He worked virtually continually on this "oil-drop" experiment from 1909 to 1912.

BREAKING THE OIL-DROP NEWS TO THE WORLD

In the fall of 1910, when he finally believed that he had all of the necessary data, he submitted his findings for publication. The physics community did not wait long to recognize his groundbreaking achievements. Even the German physicist and discoverer of X rays, Wilhelm Roentgen, who had been so skeptical earlier, now lauded Millikan's work. And the even harder-to-convince German chemist Wilhelm Ostwald of the University of Leipzig wrote in 1912: "I am now convinced. . . . Experimental evidence . . . sought in vain for hundreds and thousands of years

. . . now . . . justify the most cautious scientist in speaking of the experimental proof of the atomic theory of matter."[6]

Once one knew the electrical charge, it was possible to calculate many other things, such as the volume of air molecules in a cubic centimeter, the interatomic distances in a crystal, and the weight—in grams—of any atom whatsoever. Indeed, it can be argued that the charge on the electron is second in importance only to the velocity of light among constants in nature. The greatest significance of this work was that it forever convinced even the hardiest skeptics that the atomic theory of matter was correct. All matter in nature really did consist of atoms. (Interestingly, Millikan's "oil-drop" technique has surfaced again relatively recently: David Rank of the University of Michigan used it in the 1960s to try to actually locate quarks, perhaps the ultimate "building blocks" of nature. Atomic theory had already strongly suggested their existence, so efforts to actually find them were reasonable. To locate them, Rank tested meteors, crustacean shells, moon rocks, sediment lava, and so forth. Although Rank had no real success, physicists are still strongly inclined to believe that quarks exist.[7])

In 1921, Millikan accepted, at Hale's insistence, the directorship of the new Norman Bridge Laboratory at the California Institute of Technology. The lab officially opened on January 29, 1922. Later that year, he became the head of Caltech. His career as president of the California Institute of Technology was long and brilliant. He brought the eminent geneticist T. M. Morgan to Pasadena, as well as Robert Oppenheimer, the most brilliant of the new breed of theoretical physicists. He introduced a humanitarian and religious dimension to the Institute for the very first time when he appointed the theologian Theodore Soares to the faculty. Barely two years after his appointment to the Norman Bridge Laboratory, in 1923, Millikan captured the Nobel Prize for his work in measuring the charge on the electron, for proving the atomic theory of matter, and for providing the experimental backing for Einstein's theoretical ideas of 1905. From this point on, physics would change even faster.

DRAWBACKS OF THE CLASSICAL THEORY OF LIGHT

One of the most exciting new changes had to do with the very nature of light itself. With the main problems of the oil-drop work behind him, Millikan turned his attention to Einstein's new work on the "photoelectric effect." Newton had begun his own "revolution" in physics with the presentation in 1686 of his Principia *De Motu Corporu* (*The Motion of Bodies*) to the prestigious scientific organization, the Royal Society of London. In this he began his analysis of the laws of motion. And in his 1704 treatise, the *Optics*, he suggested the idea that light was made of particles. Although his work on motion was, of course, an approximation of the truth even today, the particle theory of light did not grab hold. For one thing, as Newton himself admitted, it could explain only reflection and refraction. According to Newton's view, reflection was merely the "bouncing" of light particles from a surface. Refraction was the bending of light when it traveled from a rarer to a denser substance, such as from air to water. A change in velocity as the corpuscles of light penetrated the denser substance in turn caused the bending. However, with other topical phenomena such as diffraction (the modulation of waves in response to an obstacle) or polarization (the phenomenon whereby light waves exhibit different properties in different directions), Newton's theory of light was helpless. It could not explain the latter because both diffraction and polarization were manifestations of the *wave*, rather than the particle theory of light. So a "particle" theory of light was useless.

Instead, Newton's contemporary, Christiaan Huygens of the University of Leyden in Holland, suggested the radical notion that light was a disturbance in the "ether." It was easy to understand the attractiveness of Huygens's conception. It was a good example of what the philosopher Wittgenstein meant by being held captive by a "picture." There was a powerful temptation to think of light waves as similar to *water* waves. And, since waves needed a medium in which to travel, i.e., water, so too did light waves need a "medium." Not seeing one, they invented it—the mysterious, ineffable "ether." The wave theory received a further boost 50 years later when James Clerk Maxwell, the Scottish physicist who first showed, in 1864, that electricity and magnetism are really merely two different aspects of the same basic phenomenon, contended convincingly that the wave idea was correct. He did this by arguing that light was an oscillation of magnetic and electrical waves—an idea that would survive for many years.

HERTZ'S WORK

Then trouble for the "wave" theory began. German physicist Heinrich Hertz, the discoverer of radio waves, realized in 1887 that ultraviolet light induced electrical charges on metal surfaces. J. J. Thomson showed further that a positive charge on the metal could be explained as resulting from the ejection of negatively charged electrons from the surface of the metal.

The problem with all of this was that it did not square with Huygens's classical "wave" theory. According to that theory, the greater the intensity of light hitting a metal surface, the more energy the ejected electrons should have; just as, with increasingly powerful winds buffeting the sea, the greater energy the water waves will have. In other words, the classical and intuitively plausible idea was that energy was supposed to vary directly with intensity. Instead, what actually happened was that with an increase in light intensity, the energy of the ejected electrons stayed the same, but the light ejected greater *numbers* of them! The only thing that increased the energy level of the ejected electrons was an increase in the *frequency* (and thus a change in color) of the light shone upon it—not its intensity.

THE PARTICLE IDEA REVIVED

So it was that in 1905, Einstein testified that this so-called "photoelectric effect" could be explained only by the idea that light was composed of *particles*. According to Einstein, light beams were composed of tiny "bundles" ("quanta") of energy. In the classical theory of Maxwell, color is identified with frequency. So, whenever light beams hit metal, a few particles of light are absorbed. In some cases, the energy absorbed by the metal surface is sufficiently energetic to free electrons from the surface atoms of the metal. Einstein expressed this relationship by an equation that linked the ejected electron's energy with the frequency of the light shone upon it. In Einstein's equation, the energy of the light quanta shone on the metal equaled frequency times a constant— identical for every color. In other words, the energy of the light varied with the frequency of that particular light beam. And so, therefore, did the energy of the evicted electron. The greater the frequency of the light shone upon the metal, the greater the energy of the evicted electrons. (The electrons do not produce the light; the light causes electrons to be emitted from the metal.)

Out of respect for Max Planck, Einstein used the symbol h, or "Planck's constant," which represents the ratio of the energy of a quantum of radiation over its frequency. In so doing, he was able to conquer theoretical problems in radiation that many physicists had believed were impossible to solve. Merely to have some kind of label for the constant, Planck had, for no special reason, written the constant as h and did not think it had any special importance—merely an ad hoc device to allow the equations to balance. But Einstein had realized that the "h" was far more than an "ad hoc" convenience; it was still one more clue to the nature of reality, as Einstein demonstrated when he explained the photo-electric effect.

And by looking at the constant in this new way, Einstein, via the photoelectric equation, brought about the first really important application of quantum theory. That is, he was able to give the correct analysis of the photoelectric effect as explained above.

MILLIKAN'S ROLE IN THE "PARTICLE" THEORY OF EINSTEIN

So far, this was all theory. But it was at this time, while working in Europe in 1912, that Millikan began to think about experiments to test these ideas. As already noted, he intended to show—and did show—that Thomson was right about the atomic structure of matter and the mass of the electron. But he also wanted a verdict on Einstein's photoelectric effect. And the Ryerson Laboratory at the University of Chicago was ideally suited to this task. In Millikan's own words:

> I knew from my Berlin discussions how difficult it would be to get a convincing answer to the problem of this Einstein photoelectric equation, . . . but the question was very vital and an answer of some sort had to be found. . . . I began this phase of my photoelectric work in October, 1912, . . . I set my pupil, Dr. Hennings, at making careful stopping-potential measurements in the case of the common elements while I went at the more difficult problem of using the alkali metals and making tests over what was up to that time an unprecedented range of wave-lengths.[8]

Was Einstein right in claiming, against Planck, that the value obtained by multiplying h times the frequency of light really did equal the energy of a single quantum?

Also, did the photoelectric equation square with experimental observations? Finally, *was* "h" the same for all colors—or merely an artifact of only a few? Millikan was enormously skeptical. If Einstein was in opposition to classical mechanics, the idea of the photon, the basic unit or "particle" of light, had to go. [It has to be noted, however, that there is considerable confusion about what Millikan thought he was testing and, more importantly, what Millikan thought that Einstein actually did or didn't believe: some historians, notably J. L. Michel (of the University of Chicago), argue forcefully that Millikan had misunderstood some of Einstein's work, often making the error of thinking that Einstein had abandoned the quantum theory. Michel points out that Millikan had incorrectly believed that Einstein's ideas were based on other ideas of German physicist Max Planck and English physicist J. J. Thomson, leading him to the above error. This led Millikan to have, for a while anyway, a negative opinion of Einstein's photoelectric work for no good reason.][9]

With his usual ingenuity, Millikan conceived of an elaborate experiment specifically designed to carry out the tests that would answer these questions. His approach was based on the earlier work of one of his students, William Kadesch. To test Einstein's claimed relationship between the energy of the ejected electrons and the light frequency, he devised his famous "vacuum" apparatus that allowed him to shine light of differing frequencies on a variety of metals. He measured the energy of the electrons coming off the metals when he hit the metals with light of varying frequency. The energy of the electrons could be measured using simple Newtonian ideas: the greater the amount of electrical force needed to stop an electron, the greater its energy must be.

Once he had determined the speed of an electron, he could easily figure out the energy of the electrons emitted from the metal surface when bombarded with a particular wavelength of light (color). Similarly, it is possible to calculate the energy of an oncoming truck if you know how fast it is moving. He went through the same procedure for every color in the electromagnetic spectrum.

EINSTEIN TRIUMPHANT

Einstein was right on all counts. "Planck's constant" really held true all through nature and he had proven that Einstein's equation and theory were entirely correct. The energy of evicted electrons did really depend on the frequency of light shone on it. Millikan was able to calculate, for example, that Planck's constant was 6.57×10^{-27} erg second. Finally, in 1916, he wrote:

> Although I have at times thought I had evidence which was irreconcilable with that equation, the longer I have worked and the more completely I have eliminated sources of error, the better has the equation been found to predict the observed results.[10]

Oddly, in a fit of fabled obstinacy, while he eventually accepted the *mathematical* utility of Einstein's equation, he continued to resist the notion that there actually were such things as quanta or "light particles" in the real world. Nobody really knows why, except that Millikan's stubbornness in the face of evidence was legendary. His resistance to abandoning his theories about cosmic rays in the face of overwhelming evidence is still another example. Often, his religious fundamentalism played some role in his unreasonable resistance to new ideas and new evidence. Historians Roger Stuewer and Robert Kargon have both pointed out, incidentally, that Millikan, in his autobiography, tries to give the impression that at the time he had accepted the reality of photons—a claim that both Stuewer and Kargon find "shocking."[11]

THE NATIONAL RESEARCH COUNCIL

Millikan had his patriotic side as well. Like Michelson, he was quick to react when his country appeared to be threatened. When war broke out in 1914, one event in particular—the murder of 27-year-old Henry Moseley at Gallipoli in August 1915—dramatically affected Millikan. Moseley was a most promising young physicist doing pioneering work on X rays. As Millikan said at the time, the murder of Moseley alone made the war "one of the most hideous and most irreparable crimes in history."[12]

By September 20, 1916, the U.S. government had begun to develop the idea of a National Research Council, in accord with the collective will of the National Academy of Sciences. Millikan was more than eager to help. Endorsed by Woodrow Wilson, the purpose of the council was to provide a tightly organized body to foster scientific research for "the national security and welfare." Thus, the council came into existence not long after the sinking of the *Lusitania*, in response to an official request by Woodrow Wilson. However, the *Lusitania* notwithstanding, the real catalyst for this move was the infamous attack on the *Sussex*, which enraged Congress, the public, and Millikan. Shortly after this attack, on April 18, 1916, Wilson sent Germany an ultimatum to either cease all hostilities or risk war; Wilson then promptly followed the Academy's suggestion and appointed the illustrious

American astronomer George Ellery Hale as the Council's head. In the Academy's own words, it strongly suggested

> that there be formed a National Research Council to bring into cooperation all existing organizations with the object of encouraging the investigation of natural phenomena, the increased use of scientific research in the development of American industries, the employment of scientific methods in strengthening the national defense. . . .

WORLD WAR I WORK

In a short time, the National Research Council would establish itself as a critically important institution, not only militarily, but for the future progress of American and world science. Soon after the armistice, it began its fellowship program and started publishing the *Bulletin of the National Research Council*. These developments were not without some controversy. There has been some discussion by historians of the "quality control" in doling out the National Research Council fellowships. In a letter of December 2, 1953, for example, John Slater of MIT complains that ". . . I had a National Research Council fellow last year who proved not to be up to the level of the rest of the group, and the year was really wasted for him. . . ."

Such inevitabilities aside, Millikan took control of the physics division of the National Research Council, while Michelson studied the problem of creating devices to detect enemy submarines.

OTHER WORK

Millikan's important work did not end with his electron and photoelectric research, even if they remained his finest achievements. He made inroads into understanding other aspects of atomic physics as well. Along with his colleague I. S. Bowen, he developed a "hot spark" technique that allowed them to study the spectra (any distribution of electromagnetic radiation, such as the colors of an ordinary rainbow) of atoms that had had all but a single electron removed. These results culminated, ultimately, in the abandonment of old Bohr-type quantum ideas, such as the belief that a fourth "quantum number" was needed to accurately describe the spin of an electron. (The first described the distance of the electron from the nucleus, the second described the very fine lines in the atomic spectrum, and the third described the magnetic qualities of an electron circling a nucleus.) In quantum mechanics, physicists often find that properties of a physical system, such as time and momentum, can take on only certain discrete numerical values; thus, if the discrete values are 2.2, 2.4, and 2.6, the system cannot have a value of 2.3 or 2.5. Because it is a principle of science to keep explanation and description as simple and economical as possible, scientists are loath to introduce new quantum numbers unless absolutely necessary. And the keeping of a fourth quantum number would have violated this principle of economy.

COSMIC RAYS

In the early part of 1932, Millikan had immersed himself in research on cosmic rays. Sadly, this turned into an acrimonious and sometimes bitter debate between Millikan and the Washington University Nobel laureate physicist Arthur Compton over the ultimate nature of these rays. However, some of the more pleasant and positive results of this conflict appear in a letter of January 8, from British physicist Joseph Boyce to British physicist and Nobel laureate John Cockcroft, a friend of Boyce's at the famed Cavendish Laboratory in England:

> . . . Some of the photographs show simultaneous ejection of (+) and (−) particles of high speed, as if both a proton and an electron were knocked out from a nucleus by the cosmic ray. . . . With that and the high voltage developments everywhere it looks as if cosmic ray work will become a laboratory problem for a while rather than a mountain-climbing excuse.[13]

Though Compton eventually proved to be right in his theories about cosmic rays, Millikan, quite unintentionally, had midwifed one of the other legendary discoveries in physics—the positive electron or "positron," long predicted by British–Swiss theoretical physicist and Nobel laureate P. A. M. Dirac's theory of the electron. Later, Anderson, in studying the passage of cosmic rays through his cloud chamber, a common laboratory device for detecting subatomic particles, had noticed something peculiar—something that behaved like an electron, but that seemed to have a positive rather than a negative charge. In August 1932, Anderson described the tracks of the "positron" as proof of a "positively-charged particle comparable in mass and in magnitude of charge with an electron."[14]

Perhaps that was some consolation for Millikan.

GROWING CONCERNS OVER ATOMIC PHYSICS

By the 1920s, criticisms of science began to escalate. Research in atomic physics was going on at the time, even if it had not yet become the dominant field it would later be. Not surprisingly, many were becoming increasingly alarmed about its dangers. Physicist H. B. Fosdick, for example, voiced his own fears in no uncertain terms in his book, *The Old Savage in the New Civilization*. Teller, for one, was disturbed by this kind of fear and would, much later on, defend the use of nuclear power in *Our Nuclear Future*.

Millikan too was disturbed. He possessed a well-defined utopian vision that consisted of equal parts of scientific wisdom and religious faith. The universe was God's handiwork and scientists were merely doing God's work. So it was that in his presidential address to the American Association for the Advancement of Science in 1929, he gave a stirring talk titled "The Alleged Sins of Science," later reprinted in *Scribner's Magazine*. Millikan argued persuasively that science was all glory for mankind, if only men would not misuse it. The fault of the abuse of science was man's, not God's. In the proper hands, science could make life easier for all. This sort of religious talk made him a pariah at Caltech, as many scientists of that period testified. Chemist Ernest Swift told me candidly, "We all just laughed at Millikan; since religion didn't follow the scientific method."

But Millikan was not intimidated. As an example of the beneficent use of science, he cited the automobile. And in 1924, he summed up his vision of science, saying that "The discoveries which I myself have seen since my graduation transcend in fundamental importance all those which the preceding 200 years brought forth."[15] He also prophesied, later the same year, that "it is in the field of biology [rather] than of physics that I myself look for the big changes in the coming century."

While the critics of science surely had valid points, from another perspective, America may have been naive about science. Physicians and physicists already knew that radium could help cancer victims, and the public bought the notion that "more was better." A 1929 pharmacological manual listed nearly 100 medicines infused with radioactivity. Allegedly, there were radioactive chocolate bars, radioactive toothpaste, and even radioactive substances to aid sexual potency.

Despite Millikan's inability to see the point of the fears raised about the misuse of science, he was still appreciative of the contributions of other scientists. In 1948, at the dedication of the Michelson Lab in California, he listed what he believed to be the most important scientific experiments of recent history. He mentioned Michelson's famous ether-drift experiments and the discovery of electromagnetic induction by Faraday. He did not list his own.

It is no exaggeration to say that Millikan midwifed most of the major 20th century developments in science. In addition to his support of Einstein's work, the discovery of the positron, and the birth of quantum theory, he played a major role in making the California Institute of Technology one of the premier centers for scientific research in the world. Small wonder that he ranks among the great scientists of history.

CHAPTER 6

The Growth of American Astronomy

THE LEGACY OF HALE

Astronomer George Ellery Hale was born in Chicago on June 29, 1868. Though he originally intended to join his family's elevator business, his basic instincts soon turned him to science.

His finest contribution to science was, arguably, the spectroheliograph, a device for taking photos of the sun using only the sun's spectral lines. Using it he took the first really outstanding photos of "sunspots"—those hugely powerful regions of solar activity. By 1908, he had discovered magnetic fields in these sunspots.

In addition to his scientific work, Hale was equally and justly renowned for his tireless fund-raising campaigns. Coming from a wealthy family of elevator manufacturers, Hale was accustomed to the ambience of money. Thus, he was not afraid to pursue such financial titans as Andrew Carnegie. Carnegie was not, of course, the first millionaire to contribute to American astronomy. Both Hooker of Pasadena and the traction magnate Yerkes had also aided the study of the heavens with substantial donations.

In 1897, Hale orchestrated the construction of the Yerkes

Observatory as well as its 40-inch telescope. Again because of Hale's unrelenting efforts, by 1917 the United States had the magnificent 100-inch telescope on Mt. Wilson (to be later surpassed by the great 200-inch telescope on Mt. Palomar).

HALE AND CALTECH

Another of his great passions was Caltech. Its predecessor, the old Throop Institute, had been singularly undistinguished. But under Hale's guidance, Throop evolved into the California Institute of Technology. According to historian Raymond Fosdick, Hale had hoped "this titanic tool of science might bring into fresh focus . . . the universe, its apparent order, its beauty, its power." Soon, via promises of ample laboratory space and money, Hale was able to attract such luminaries as physicist Robert A. Millikan and chemist A. A. Noyes.

HALE AND THE NATIONAL RESEARCH COUNCIL

Still another of Hale's passions was the National Research Council (NRC). Founded in 1917, the Council helped organize science for national defense, without any intrusions from the U.S. government, conducting its research in such places as the laboratories of the Western Electric Company. The Council would later provide grants for a very wide variety of projects in science, including some of the earliest important work done with the electron microscope (a microscope far more powerful than the ordinary optical microscopes seen in any average college laboratory) by Linus Pauling, R. T. Birge, and many others.

Hale died of pneumonia in Pasadena, three months short of his 70th birthday. In his will he offered a permanent financial legacy to the Institute:

> for providing the salary payments of research fellows, one or two a
> year, throughout a period of years. . . . These research fellowships at

the said California Institute of Technology shall be known as the George Ellery Hale Research Fellowships in Radiation chemistry. . . .

THE STARS

The work of Hale and his pioneering studies of the sun's physical and chemical properties at the California Institute of Technology revolutionized our knowledge of the sun. But, like so many acquisitions of new knowledge, they only pointed to new doors yet to be opened. By the beginning of the 20th century, scientists had already amassed a vast amount of information about the sun. They documented that the sun was 93,000,000 miles from Earth and that its light took 8 minutes to make the journey here. Its diameter was 864,000 miles and its mass the equivalent of about 331,950 Earths. They deduced that its temperature farthest from the surface was 2000°C, rising to 6000°C in the flames themselves and far higher than this in the interior of the sun.

They had also acquired reams of information about the chemistry of the sun. But the sun was merely a medium-sized star. What about the others? Not surprisingly, the information gathered about the sun would be a powerful clue to the nature of other stars.

Yet the accumulation of such knowledge is a long and difficult process. The exact mechanism of energy production in the sun still remained a mystery for future scientific theorists to work out. (Hans Bethe of Cornell, much later on, would make a remarkable contribution to this problem.)

A DOOMED UNIVERSE?

Then another kind of question naturally arose: What was the ultimate explanation for the existence of stars? What was the primal force that created the cosmos? For these questions, science alone could not do the job. Scientists turned to philosophers and philosophical theorizing, as they still do today. For the explanations of such ultimate cosmological questions, theorists could, both then and now, do no better than to begin with the scholastic arguments for the existence of some kind of deity.

Science would do better with questions about the future of the stars and, indeed, the future of the universe itself. One critical issue scientists debate today is whether the universe will ultimately come crashing down on itself, or keep on expanding forever. Today's astrophysicists know exactly what additional information they need to know in order to answer such questions about the future of the stars and the cosmos: they have to know how much mass is in the universe, since the more mass, the stronger the gravitational forces at work, and the more likely that the universe is to someday collapse. For example, many astrophysicists believe that there is much "dark" or invisible matter in the cosmos. One interesting possibility, for example, is that *neutrinos*, the subatomic energy-bearing particles in the atom, may have some finite mass, albeit extremely small. If that is true, then neutrinos would provide tremendous amounts of mass beyond what is already known. Black holes are another possible source of "invisible" mass.

BESSEL'S WORK ON STARS

It was most likely the 19th century German astronomer F. W. Bessel who conducted the earliest important studies on stars. In 1818 he published his "Fundamenta Astronomiae," which catalogued over 3000 stars. By 1838, he had studied the shift in position of a star comparatively "close" to the Earth—Cygni 61, which is just barely visible to the eye. In so doing, he was able to show that it was 40 trillion miles away. This was, in effect, the first parallax measurement of a star. (Parallax is a distance-estimating technique based on the fact that the locations of things appear different when viewed from different angles. The amount of difference can, in turn, be used to estimate its distance.) Yet

Bessel's method involved enormously complicated trigonometry. Without the benefit of modern computers, the calculations took months. Even in the early decades of the 20th century, scientists had determined the position of only about 60 stars. Later, of course, with more efficient techniques and computers, the work went much faster.

The remarkable thing about Bessel's work was that even though Cygni was "close," comparatively speaking, it was still an unimaginably large distance away, so that any change in its position had to be minuscule. In Cygni's case, it was "close" enough that this change in position, though very small, was still just barely measurable. For any stars substantially farther away than Cygni, however, we simply would have been unable to see any apparent change in position. Measuring the change in Cygni was, in order of difficulty, like trying to notice how much the grass grows in a period of about five hours.

Another approach relied on the fact that the *motion* of a star was a clue to its distance: the closer the star is, the more it appears to be moving relative to us. But this method too had its limitations, since for a star at a very great distance, we simply would not be able to detect any apparent motion relative to us.

The road to measuring greater stellar distances, therefore, required other techniques. Even the nearest Cepheid star (so called because astronomers found the first ones in the constellation Cepheus), the Pole Star, was much too far to measure by parallax or any other known method. Astronomy might have stopped right there had it not been for the fact that Henrietta Leavitt found another method.

LEAVITT'S WORK

In 1912, one peculiar batch of spectrographic "photos" appeared to show something quite unbelievable—a particular star seemed to be alternately growing closer and farther from the Earth. In other words, it alternately brightened and then dimmed.

What could possibly explain the observation that the source appeared to be both coming closer *and* moving farther away from us?

Many scientists supposed, naturally enough, that they were really seeing two stars close together. If one were brighter than its companion, and both were constantly moving around one another, that could explain things neatly. But the mathematics of the theory did not work out and the idea had to be jettisoned.

At this point, Henrietta Leavitt, a student of Pickering's and a graduate of Radcliffe, tackled the problem. Her work told her that there was a collection of such "Cepheids" in the extremely distant cluster—the Lesser Magellanic Cloud—a mighty star group named after the explorer Ferdinand Magellan, most visible in the southern hemisphere.

Working at the Harvard University Observatory in the town of Arequipa in Peru in 1912, she immediately rejected as unworkable both the idea that there were really "two stars" and the idea that the same star was alternately waltzing closer to and farther from us. She then worked out her new approach. Studying the Cepheids in the Magellanic Cloud, she saw, as did previous astronomers, that these oddly behaving stars would vary in their luminosity over a period of time. But she saw still further than other astronomers and noticed that the brightest Cepheids remained bright for longer periods than the fainter ones. Some stars might have rhythmic pulsation patterns that were scarcely more than a day, while others had periods lasting up to four months.

Thus, from the period of brightness of a Cepheid, she realized that she could read directly its actual luminosity and from its luminosity, the period could be read. She thereby constructed a "period–luminosity" curve, which plotted the period of the star against the absolute brightness. The graph showed clearly that the brighter a star, the longer the time between pulsations and vice versa.

But, a century before this work, the great Enlightenment astronomer Count Rumford had noticed still *another* factor related to brightness. He saw that the luminosity of an object was, in turn, *also* a function of the square of its *distance*. Thus, one light bulb three times as far away as another of equal brightness would appear nine times less bright.

What Leavitt did was collate all of these ideas. Working with Danish astronomer Ejnar Hertzsprung,[1] she realized that the period–luminosity idea could be coupled with Rumford's discovery. In other words, the luminosity could also be used to measure distances of astonishingly distant stars—distances much too vast to be measured by traditional methods like parallax.

However, she needed a reference point—a star or group of stars whose absolute brightness and distance was known. The brightness of other stars could then be measured against the reference stars. As she discovered, the Cepheids in the Magellanic Cloud could be used as just such a "reference" point or "standard." The reason they served so well in this capacity was that these Cepheids were close enough that their distance and, therefore, their actual brightness could be easily measured using ordinary methods. With this information, stars that were too far to be measured by parallax could still be measured indirectly by Leavitt's method.

All that was then needed was to study the elusive star and measure its pulsation period. Once she knew that, she used her period–luminosity curve to determine its actual brightness. Finally, by measuring the difference between the actual and apparent brightness, she could determine the star's distance from us. Now here is where the Cepheids shine to great advantage. Those stars whose apparent brightness was *less* than the Cepheids could be safely assumed to be farther from us, those that were brighter were closer. In more quantitative terms, if, for example, one Cepheid appeared nine times as bright as another, it must be three times closer to us, since luminosity is a function of the square of the distance. In this way, astronomers now had a method for estimating virtually any stellar distance.

Leavitt was one of America's and the world's most distinguished astronomers and worked for many years studying photographs of the heavens to determine the composition and distribution of stars. She would go on to discover 25 Cepheids in the group.

HARLOW SHAPLEY

Subsequently, the biologist and newspaperman-turned-astronomer, Harlow Shapley, refined and developed Leavitt's system. Shapley was of the few scientists in American history who began achieving truly remarkable results at an early age. Dutch astronomer J. C. Kapteyn, in a letter to Hale of February 7, 1919, described him as a ". . . brilliant man."[2] He was to become the next director of the Harvard Observatory, after the death of Pickering in 1919.

According to Hale, Shapley was one of the most daring of the younger scientists. In the manner of Linus Pauling, Einstein, and Oppenheimer, he was, according to Hale in a March 29 letter to President Lowell of Harvard, ". . . much more venturesome than other members of our staff and more willing to base far-reaching conclusions on rather slender data. . . ."[3] And Lowell meant this as a compliment, for there can be such a thing as too much experimentation and too much data collecting. That was the great fault of solid-state physicist John Slater a few decades later, for example. While careful quantitative work is of course critically important, good science requires *some* risk taking and speculative theorizing as well.

Perhaps the nadir of Shapley's scientific career was a public debate regarding sizes and dimensions of scale in the universe with American astronomer Heber D. Curtis, known for his determination of the distance of the Andromeda Galaxy. According to Owen Gingerich of the Harvard–Smithsonian Center for Astrophysics, Shapley, only 30 at the time of the debate, was "clearly nervous at the prospect of facing an experienced public speaker . . ." Gingerich goes on to point out that the debate was a catastrophe for Shapley, inasmuch as Curtis displayed more maturity as well as knowledge.[4]

Nevertheless, Shapley made many stunning contributions to astronomy. Among his early work was a variant of Leavitt's idea. He discussed the Cepheid stars and pointed out that the length of the period of these stars varied directly with their redness. In this theory he also suggested the now-familiar idea that stars contract over time, thereby becoming bluer. Based on visual observation of these factors, he was able to give rough estimates of the dimensions of the Milky Way: It had a radius of 10,000 light-years, though at its "thickest" point, it measured only 1000 light-years. He therefore concluded, correctly, that it was a flat "disk" of stars.

Working with the 60-inch telescope at the Mt. Wilson Observatory, he focused on Cepheids grouped near the sun—the so-called "globular clusters." Having realized through Leavitt's work that such clusters were important, he began using Cepheids in our own region of the cosmos as his reference point. He then determined, based on the observations of the same factors mentioned above, that they were located in the center of our galaxy.

By 1917, he had further shown that many of the different clusters were as far as 200,000 light-years. The farthest was NGC 7006—about 220,000 light-years away. Also, he proved that the cluster of Hercules had a population of about 35,000 stars and was about 36,000 light-years away. With still more mathematics coupled with adroit scanning of the heavens, Shapley was able to prove that this cluster was about 50,000 light-years away, in line with the constellation Sagittarius.

Perhaps most importantly, these determinations proved that our solar system was not "the center of the universe" as earlier astronomers like Herschel[5] had believed. In reality, we were negligently placed somewhere on the outskirts of our galaxy.

HUBBLE AND THE MT. WILSON TELESCOPE

Later, in 1924, the fabled Edwin Hubble of the Mt. Wilson Observatory used the great new telescope at Mt. Wilson to probe the remotest districts of astronomy—unimaginable distances that no one ever dreamt could really be studied.

Edwin P. Hubble entered the world in 1889 in Marshfield, Missouri, though he spent most of his formative years in Kentucky. Early on, Robert Millikan, then at the height of his powers in the early decades of the 20th century, noticed Hubble's talent in physics and offered him a scholarship to the University of Chicago where Millikan was then teaching.

Curiously, Millikan and Hale probably saved Hubble from a disastrous (from the point of view of science) career as a prize fighter. Boxing had been an avocation for Hubble early in life and during his college years in Chicago, a sports promoter even wanted to set up a match between Hubble and the world's heavyweight boxing champion, Jack Johnson. This idea was not as crazy then as it may sound today. It seems Hubble did have athletic talent, and the opportunities and money for American athletes at the time were increasing rapidly. Jack Johnson was, after all, the world's first black heavyweight champion.

Fortunately for science, Hale and Millikan dissuaded him and he headed for Oxford as a Rhodes scholar instead. Still having trouble making a career decision, Hubble studied law at Oxford and even practiced for a while in the United States. Though his graduate training was in law rather than science, his early interest in physics and astronomy nagged at him and he finally took a post at the Yerkes Observatory in Wisconsin. In 1916, he wrote a paper titled "A Photographic Investigation of Faint Nebulae." When he finally came to Pasadena, he began his study of the thousands of already-discovered nebulas, or large clouds of gas and dust in interstellar space, including Messier 31, the spiral nebula in Andromeda, and Messier 33, a nebula in Triangulum.

Soon, he turned the new telescope on Mt. Wilson on the Andromeda nebula. It proved to be powerful enough to resolve into individual stars what were previously only faint shadows on Andromeda's rim. With the improved view, he began to suspect that Andromeda might constitute a *second* nebula, in addition to our own Milky Way. Using the techniques described earlier, he proved that the great Andromeda nebula was 900,000 light-years away—far beyond our own galaxy. For that reason, it comes as no shock to learn that it is only a faint patch of light in the night sky.

Soon, Hubble and his associates had resolved hundreds of nebulas. The newly discovered gases, clouds, and other stellar debris were as Herschel deemed them, "primeval chaos of shining fluid." Perhaps most startling, having estimated the distance of the Andromeda Galaxy, Hubble proved that the size of the universe was *twice* what science had previously thought it to be. It was also much older.

THE EXPANDING UNIVERSE

Finally, in the mid-1920s, Hubble worked out his greatest and most mutinous fantasy: he positively convulsed the astronomical community when he theorized that every star in the firmament is moving away from us, from which he concluded that we live in an exploding and expanding universe. (This is about the only conclusion possible, save for the "option" of rejecting the idea that the speed of light is not the constant scientists had previously thought it to be; the speed of light is, of course, established beyond any reasonable doubt.) As Hubble and his aid Milton Humason surmised at the time, the universe would double its diameter every 1400 million years with this kind of expansion. The consequence was "Hubble's law," which states that the farther a galaxy is from us, the faster it is moving away from us.

He made the measurement, as is now well known, by studying the "redshift," the phenomenon whereby the wavelength of light increases as an object moves farther away from us. Anything moving away from an observer would shift its wavelength to the red end of the spectrum.

The first to apply this technique to measure the speed of approaching or receding light sources was the amateur astronomer William Huggins. Hubble further noticed that the speed of recession of a nebula increased as the nebula increased its distance from the Earth. Another and even more striking result of redshift studies is the fact that the farther the galaxies are from us, the faster the galaxies are receding. As it turns out, galaxies at a distance of about 6 billion light-years are moving slightly faster than half the speed of light.

Soon, new information buttressed new speculations about the expansion of the universe. In 1959, Austrian physicist Hermann Bondi and British scientist R. A. Lyttleton suggested that the charge on the proton is infinitesimally greater than the charge on an electron. Although the difference in charge between individual subatomic particles is unfathomably small, it adds up to a significant amount when one considers the awesome total numbers of protons and electrons in the universe. So, in the universe as a whole, the net *total* positive charge would be slightly greater than the total negative charge on the electrons. Thus, the entire universe would have a net positive charge. But since like charges repel, all stars and galaxies would be constantly repelling each other, thereby "pushing" one another farther away and easily accounting for the expansion of the universe.

By extrapolation, then, scientists suggest that between 10 and 20 billion years ago, the entire universe was squeezed into a region of space about the size of a hydrogen nucleus. In that state the density of the universe would have been about that of one of the larger planets in our solar system. Then, this fantastically compressed mass destroyed itself in a cataclysmic explosion. Though scientists do not know why such an explosion occurred, British physicist Stephen Hawking has suggested that it may be connected with the gravitational collapse of a star into a "singularity," or a compression of matter into a region of zero volume plus the possibility of "time-reversal," so that the contraction suddenly becomes an explosive *expansion*, though Hawking himself now is not so sure of this explanation. Many scientists thus believe that the galaxies we see moving away from us are merely the fragments of this original "big bang"—hence the name of this theory, first published in 1927.

DIFFICULTIES FOR THE BIG BANG THEORY?

Evidence for the theory has been changing rapidly in recent months. In studies carried out recently by powerful X-ray and infrared telescopes, astronomers have discovered large groups of galaxies, centered around quasars—the great concentrations of energy at the center of many of these galaxies—at a distance of about 10 billion light-years from Earth. The galaxies, in turn, included at least one fantastic entity called the "great wall"—a line of stars at least half a billion light-years in width. These discoveries indicate that such enormous galaxies may have formed only a few billion years after the original big bang. The problem for the big bang theory therefore is that, if the above observations are correct, the time scale predictions may be wrong. For according to current theory, the universe must be about 15 billion years old. Yet the above findings suggest that the universe is nowhere near that old.

On the other hand, most recently and as of this writing, astronomers are ecstatic about the discovery, announced in April 1992, of minuscule variations in temperature in the radiation left over from the big bang. The reason this is so critical for the big bang theory is that astrophysicists believe that such variations were necessary to account for the great irregularity in the universe today, with its vast regions of empty space punctuated by dense galaxies. According to Michael Turner, professor of astronomy at the University of Chicago, "The Holy Grail has been found." Interestingly, current cosmological theory had already predicted these variations in temperature, although until now, no one had found them.

THE STEADY STATE THEORY: THEOLOGY IN PHYSICS

For many years, the big bang theory was not the only viable theory. In 1948, British astronomer Hermann Bondi as well as

American physicist Thomas Gold proposed a rather different view—the so-called "steady state" theory which the popular British astrophysicist Sir Fred Hoyle popularized. While these scientists accepted the idea that galaxies are receding from us, they also assumed that the universe is even and uniform both in time and in space. No matter from what point you view it, the universe will always look the same in their theory. Later observations confirmed the uniformity of space concept.

The idea of uniformity in time, however, was more difficult. Their claim was rather fantastic: no matter how far back or how far ahead in time one traveled, the universe would appear exactly the same.

How could this be, if experiments showed that the galaxies were receding from us and, therefore, that the cosmos is expanding? Their proposed solution to this difficulty was hardly less than theological. They suggested that as galaxies move away from each other, new ones are continually being formed, thereby maintaining the overall density of the cosmos. Nature is literally and constantly creating matter out of nothing! Specifically, they suggested that if one assumes that nature creates a single hydrogen atom in a volume of space of about one cubic kilometer every year, the creation of this new matter would exactly balance the "thinning" of the universe caused by its expansion. Thus, the universe would remain in a "steady state."

For many, however, the strongest objections to the "steady state" theory were the many violations of conservation laws implied by the idea that matter appeared *ex nihilo*. At the very least, that violates the law of conservation of energy, which says that matter (equivalent to energy by Einstein's famous equation $E=mc^2$) can neither be created nor destroyed.

Nowadays, however, this violation is not so serious, since the work of Yang and Lee on parity conservation shows that the conservation laws are not so sacrosanct as once thought. In any case, Hoyle tried to answer the conservation of energy problem by suggesting that the energy for the creation of this new matter might come from the general energy of expansion of the universe itself. Although this would slow the rate of expansion of the universe, it would provide a source for the missing energy and thus preserve the conservation laws.

The real problem with the steady state theory appeared later through the work of Arno Penzias of Bell Labs, Martin Ryle of Cambridge University, and Robert Wilson of Bell Labs. Part of this new evidence was Penzias and Wilson's discovery of radiation from the early, extremely hot stages of the universe, found in 1965. That, in turn, proved that the principle of uniformity was untenable, since in the distant past, the universe must have been much denser than it is now. (The kind of microwave radiation Penzias *et al.* found could only have come from a much denser universe.) Thus, the steady state theory had to be abandoned. Whether it will remain in the museum-heap of brilliant but flawed hypotheses is another question.

WALTER ADAMS

In an earlier period, U.S. astronomer Walter S. Adams found an accurate way of measuring distances. As director of the Mt. Wilson Observatory during the years of WWI, he worked out a method of stellar parallax based on some revealing lines on a spectrographic photo, which varied directly with the temperature of the subject of the photograph.

In this way, he could estimate the brightness of the star and, therefore, its distance. In the few years before 1921, Adams and his co-workers, using this technique, worked out the distances of 2000 stars.

OTHER SCIENTISTS AT MT. WILSON

Many other astronomers worked at Mt. Wilson during the opening decades of the 20th century. Another deserving of mention is E. E. Barnard, a Tennessee photographer who charted several other areas of the night sky which were inaccessible to observers before the Mt. Wilson telescope. He also discovered the fifth satellite of the giant planet Jupiter.

Yet more fantastic revelations emerged with the Mt. Wilson telescope. For years, astronomers had been studying the star Sirius. As early as 1844, the astronomer Bessel had concluded that Sirius's orbit was an ellipse. But that demanded an explanation, since the laws of physics and the combined gravitational pull of the known surrounding planets could not account for this motion—unless there was some other, undiscovered star near it.

Because this unknown star had been so difficult to find, astronomers at the time assumed, naturally enough, that the star was not very luminous. It was not until 1872 that Alvan Clark of Fall River, Massachusetts, the builder of the 26-inch telescope for the U.S. Naval Observatory,[6] and his son Alvan Graham Clark actually found this "hidden" companion of Sirius, or "Sirius B," as they called it, a white dwarf star. It was a very faint star nearing the end of its lifetime. Not only was Clark a fine scientist, but he was one of the great builders of many of the best U.S. telescopes; in 1877, Asaph Hall discovered a moon of Mars with Clark's Naval Observatory scope. In 1881, Clark produced the 36-inch telescope for the Lick Observatory, where James Keeler made pioneering studies on the nature of the rings of Saturn.

In 1914, however, Walter Adams, again at Mt. Wilson, studied this companion star some more using the spectroheliograph, a special device for taking photos of the sun, invented by Caltech astronomer George E. Hale. Astonishingly, he found that it was actually white and extremely bright. But the light that appeared on Earth was far dimmer than the sun's, making it an object with a diameter only 1/19th that of the sun, or slightly larger than the planet Venus. That kind of brightness together with that small a size allowed for only one possible conclusion: it was unimaginably dense—denser than any object anyone had ever studied. Every cubic inch weighed many, many tons. (This was a faint anticipation of the 20th century phenomenon of supercompact "neutron" stars.)

The work done by the brilliant men and women at the great observatories like Mt. Wilson, Mt. Palomar, Kitt Peak, and many others was vast in scope and prodigious in importance. Soon, there was little skepticism as to the wisdom of Hale's investing such enormous sums of money into institutions like these. And little could any of them dream that even greater wonders lay dormant in the heavens awaiting our discovery.

CHAPTER 7

Irving Langmuir and the General Electric Laboratories

Irving Langmuir was one of the first of a new race—the industrial scientist who conducted "pure" research, research that was unadulterated by practical demands for improving a company's products or profits.

EARLY LIFE

Irving Langmuir was born in 1881 in the peaceful and picturesque city of Brooklyn when it was scarcely more than a congerie of cow pastures, with wood-frame houses occasionally dotting the landscape. Trains and carriages moved the people quietly from Brooklyn to Queens. The subway was nonexistent and urban crime and crowding lay far off in the future.

His father, who could have been the prototype for Mr. Micawber in *David Copperfield*, was forever in debt. But he had faith in his son's academic ability, even though he was doing poorly in school. In fact, when Irving was 11, his performance in school was so poor that his father tried briefly to have him educated in Parisian schools. The entire family had moved to Paris at that point, in order for the older son, Arthur, to study science in Europe. Education in Europe, at both lower and university levels, was, with some reason, considered superior to U.S. education.

The change of schools did not help. In fact, Langmuir simply hated school as well as discipline. He wanted to work, but could not stand being told what to do. On many occasions, instead of going to class, the young Langmuir preferred to climb mountains, ski, or go to the movies. His father soon came to the realization that he might finally flourish if he were merely left alone.

Soon, Langmuir began his love affair with science. According to some historians, he was so moved watching the funeral procession of the great biologist Louis Pasteur that his interest in science was aroused then and there. (As with most scientists, however, the interest probably would have surfaced soon anyway.) A master politician even in those years, he persuaded his father to let him attend the Chestnut Hill Academy, a prestigious private high school in Philadelphia, whose reputation in science was unequaled.

But Philadelphia was no different than Paris. His mind continued to work the same way—on its own and on Langmuir's own terms. Despite his difficult personality, he soon outdistanced his older brother, absorbing awesome amounts of information about mathematics, physics, and chemistry. According to his mother, "The child gets beside himself with enthusiasm, and shows such intelligence on the subject that it fairly frightens one."[1] Nonetheless, he was apparently not given to vanity (and about the only time he displayed any was on the infrequent occasions when he beat his wife at anagrams).

So great was his ability that he enrolled in the famed Pratt Institute in Brooklyn at the unusually young age of 14, and at 17 he was admitted to the Columbia School of Mines. In his spare time, he devoured a book on calculus, allegedly in six weeks, moving on quickly to master something a bit more challenging. Finally, he reached the educational summit in 1906 when he received a Ph.D. in chemistry from the University of Göttingen at the age of 25.

EARLY TEACHING AND FIRST CONTACTS WITH GE

Wishing to remain on the east coast of America, he accepted a position at the Stevens Institute of Hoboken as an instructor in chemistry. Though his years there were productive and satisfying, his life took a different turn after his third year at Stevens. He was not comfortable in an academic environment. So, instead of taking his usual mountain vacation, he decided to spend the summer at the General Electric corporation in Schenectady, New York, speculating that perhaps that kind of environment would suit him better. He knew that their new laboratories were carrying on important research—research that dovetailed admirably with his own interests in chemistry. At that period, Langmuir was already becoming interested in how atoms bonded together to form molecules, and in the behavior of molecules in liquids, as well as in theoretical and practical questions about the new Edison "light bulb." Because of the eventual great success of his research into these issues, he would later win the Nobel Prize. He decided to approach GE in 1909.

Langmuir's first trip to the new laboratories in the summer of 1909 was colored by some uncertainty. He knew of them only by reputation and the considerable hype fostered by GE. Such labs were unknown quantities and he could not be certain that he would find either financial security or the same degree of freedom universities traditionally offered. It was logical to suppose that he would simply be told to improve an existing product, or find new uses for old ones and so forth.

But as it turned out, he was delighted. As he said later in life, "When I joined the laboratory, I found that there was more 'academic freedom' than I had ever encountered in any university." And on the rare occasions when someone asked why he was doing this or that piece of research, his answer was always the same, "For the fun of it."[2]

His delight was in large measure due to the flexibility of Willis Whitney, the director of the GE Laboratories at this time. He allowed and even encouraged Langmuir to take as much time as he wanted studying the laboratories.

Something must be said about the laboratories themselves, given that Langmuir's name is—correctly enough—virtually inseparable from industrial scientific research. Though Edison had perhaps created the prototype of a nonacademic laboratory, it was only after the turn of the century that such laboratories became an integral part of American science. Part of Langmuir's success as well as the success of the labs was the great enthusiasm and support that GE offered to "in-house" scientific research.

The primary reasons for the growth of this type of laboratory were economic: The American business environment had, in the past 20 or so years, been experiencing profound changes, including very rapid growth, an influx of much better trained scientific personnel, and the growing power of labor unions. Yet not all of these changes were for the better. Scarcely anyone could have forgotten the Panic of 1907, for example, which was cauterized only when GE dictator J. P. Morgan imported millions of dollars in gold from Europe.

Over its history, the GE corporate moguls had been able to maintain their position of prominence and solvency by relying on the collective wisdom of the dedicated scientists of the 19th century. Still, with increased industrial competition throughout the world, it had become necessary for corporations to develop new ideas in-house to maintain a competitive edge. GE executives A. G. Davis and E. W. Rice solved this problem when they created not only new research, but helped entrench a still relatively new research concept—a university-type laboratory in an industrial setting.

Once the decision had been reached, the company lured Willis R. Whitney from the Massachusetts Institute of Technology to head their new laboratory. Whitney was a scientist with vast contacts and, during his tenure at MIT, he had befriended A. A. Noyes, George E. Hale, and George Eastman. Whitney was at the helm of the GE laboratories from 1900 to 1932. Beginning in 1928 he was also vice president in charge of research.[3]

By 1915, there were over 300 industrial research labs in the United States. By the late 1920s, there was almost a 200% increase in scientists working in such settings. In large part this was due to research in industrial chemistry, much of it fed by war-related work. At Du Pont, for example, there were nearly 2500 people working in labs by 1938, compared with a mere 850 in 1927.

Another factor important in the growth of industrial chemistry was the passage of the Pure Food and Drug Act of 1906, when it became critically important for the government to have qualified people monitoring the safety of these products.[4]

Arthur Compton, soon to be a star in the scientific firmament because of his work on cosmic rays and the "Compton effect," which described how light interacted with various kinds of radiation, would give the lab his imprimatur. He both toured and acted as a consultant to the GE lab for many years. When Compton spoke in 1918 of the great new centers for inquiry early in the century, he commented on the quality of the GE laboratories:

> It is true that the large majority of the men were working at problems that belonged to engineering rather than to pure science; but the half-dozen men who were working in pure science had, from the by-products of their work, made improvements and inventions that paid many times over for the whole expense of running the research laboratory.[5]

The summer proved to Langmuir that GE suited him perfectly and, upon receiving a permanent offer, he resigned his post at Stevens.

A RISING STAR

In 1916, Langmuir was ascending to fame; on December 27, 1916, he attended a conference on the structure of matter held jointly with the American Chemical Society, the American Physical Society, and the American Association for the Advancement of Science. G. N. Lewis, the legendary University of California chemist known for his studies on how atoms bind together to form molecules, was there, speaking about his idea that electrons were distributed around the nucleus of an atom in a cubical arrangement, with the electrons at the vertices. (Later, Lewis would be honored by being invited to publish a paper in the inaugural issue of the new *Journal of Chemical Physics* in 1933.) For Langmuir to be invited to speak in such august company was a very great distinction.

MOVING INTO OTHER AREAS OF SCIENCE

Even in his very first summer at GE, Whitney noted Langmuir's talent and energy, as the young Langmuir was already beginning to extend Lewis's ideas on how atoms combine. Also, he had begun to make new forays into the study of the factors that affect the speed and intensity of chemical reactions. Not surprisingly, GE cheerfully supported Langmuir's indulgences. By the mid-1950s, the GE laboratories had made the company even more powerful and prestigious. According to historian of science Daniel Kevles,[6] eventually, research, both at GE and at other industrial labs, would extend into many other areas of science, including the study of solid-state physics, problems in superconductivity (the ability of some substances to lose all resistance to electrical currents under very low temperatures), semiconductivity (substances with greater ability to conduct electricity than insulators, but less than true conductors), the study of crystals by X-ray and electron diffraction, and so forth.

Upon Langmuir's receiving a $10,000 award from *Popular Science* magazine in the early 1930s, his mentor Willis Whitney said of him, "Langmuir's greatest contributions to science have . . . come from his having extended his picture of matter according to the coldly logical requirements of growing theory. But he is the last man to expect to see finis written over any fundamental phenomenon."[7] And a fine example of this was his determined contributions to, and defense of, the model of the structure of proteins devised by chemist Dorothy Wrinch. Despite the fact that the

Wrinch model was scorned by most of the scientific community, Langmuir risked his reputation because he believed Wrinch's model fit the logical requirements regarding how a protein would have to be built to play the role it does in forming body tissues. In fact, Wrinch's model is regarded as having some value even today, even though the alternative "chain link" model worked out by Linus Pauling has supplanted her's generally.[8]

LEISURE IN SCHENECTADY

Schenectady was not just a place for Langmuir to work. Throughout his years at General Electric, he camped and skied in the Adirondack and Catskill mountains. A believer in nature, he worked throughout his life to preserve the wilderness and forests of New York State. And he could never separate his scientific curiosity from his allegiance to the wilderness. On one occasion, he noticed that the ice on Lake George had precipitously and rather mysteriously disappeared, and pursued the problem until he found the answer.

STUDIES ON TUNGSTEN

During Langmuir's early years at GE, their resident X-ray expert, W. D. Coolidge, had been having difficulty finding a good wire filament for the light bulb. Langmuir accepted the challenge. Tungsten itself was—or should have been—a good filament for incandescent lamps. It could be heated to extremely high temperatures, 3100°C, and therefore gave off a far more brilliant light than any other metal.

But the problem was that it kept disintegrating prematurely. Langmuir guessed that the environment of the bulb might be the trouble. Since he had done his doctoral thesis on gases, he theorized that part of the problem of the filament collapse might be traceable to the gas that routinely became trapped in the tungsten during the manufacture of tungsten. His mentor at GE, Willis Whitney, later described Langmuir's approach: "Observing, studying what takes place at the surface of hot tungsten filaments in vacuum, he introduced infinitesimal quantities of different gases. . . ."[9] Soon, Langmuir formed a hypothesis: if gas were "leaking" off the tungsten while lit, that would confirm his suspicions that gas had in fact become trapped during the manufacturing process. To test this, he tried connecting a tungsten filament lamp to the McLeod gauge. The latter was one of GE's most highly sensitive pressure gauges and with it, Langmuir hoped to see if gas accumulated in a lamp connected to alternating current. If it did, that would be at least a partial verification of his theory. The results were scarcely short of incredible. There was so much hydrogen gas emanating from the filament that it could not possibly be explained by the wire alone: in a very short time, the bulb had accumulated *7000* times more gas than could possibly be contained in the filament.

There had to be another source. He originally thought that he could produce a better bulb merely by producing a better vacuum. The reason behind his thinking was that heated glass can emit water which, in turn, can combine chemically with the tungsten filament to yield hydrogen gas. In other words, any heated glass, including light bulb glass, will evaporate water during heating, which then reacts chemically with the tungsten to give off hydrogen gas. That was, possibly the other source of gas. But in a vacuum, he theorized, this particular chemical reaction would not take place and the filament would remain intact.

But after four years of digging, he concluded that his "vacuum hypothesis" had to be abandoned: despite much trial-and-error experimentation, removing more gas was simply not yielding a better bulb. Ironically, further research proved to him that the cure for this consisted not in creating a better vacuum, but rather the *reverse*. By trial and error, he discovered that a bulb filled with nitrogen gas would actually produce a stronger and better incandescent lamp, as nitrogen acted to prevent the filament from disintegrating. This alone improved the light bulb considerably.

But there was a second, independent factor that enhanced the bulb. This second discovery came about quite by accident. When he was testing tungsten for its electron-emitting powers, he happened to pick up one that had been prepared by his colleague Coolidge for a different and unconnected purpose. He soon found that the filament could emit electrons in almost unbelievable numbers. He soon found that this tungsten was unique in having been coated with *thoria*. After more trial and error, he discovered that the filament reached its maximum efficiency when the thoria coating was approximately the thickness of a single molecule. At that thickness, the light from the filament was the most intense. (A coating of thoria that was *too* thick, Langmuir theorized, would begin to retard the emission of electrons and thus diminish the lamp's intensity.) These breakthroughs alone saved millions of dollars in electricity throughout the entire nation.

LANGMUIR IMPROVES THE DE FOREST TRIODE

The principle Langmuir discovered ranged, of course, far beyond the ordinary light bulb. The De Forest triode, for example, to date the best device for receiving radio signals, created by American inventor Lee De Forest, was in many ways similar to an ordinary light bulb. It was similar in size and also consisted of tiny filaments encased in ordinary glass. So, he reasoned that if nitrogen gas could improve the efficiency of ordinary light bulb filaments, there was every reason to suppose that the principles could be applied elsewhere. With a new and improved vacuum pump, he managed to produce a vacuum only one-billionth the air pressure of the atmosphere at sea level.

With that tube, the radio era finally emerged in America. The De Forest device was now good enough that radio signals could be clearly picked up by low-cost radios available in any department store. Eventually, the laboratory, using the results of his work, produced a tremendous variety of vacuum tubes. Some had minuscule currents, measuring mere microamperes of electricity, while others produced huge volumes of current in tubes several feet long, for use in industrial settings.

By this time, the size of the laboratory had increased tremendously. In 1907, there were less than 100 scientists. But now, barely ten years later, the GE research team numbered a solid 300.

WORK ON SURFACE CHEMISTRY

Yet Langmuir's real scientific interests had still not surfaced. Since his Paris days he had been enthralled by the "pure" science of chemistry, and it was to this great branch of science that he soon turned his attention. By 1909, Langmuir began to accept the idea that atoms and molecules were real, even forming hypotheses using these concepts. He said, for example, "The chemical activity of any solid surface depends upon the nature of, the arrangement of and the spacings of the atoms forming the surface level, and there is a close relation between the chemical activity of a surface and the electron emission from it."[10] But entrenched ideas die hard and others resisted the new concepts completely.

Undaunted, he began planning experiments that would prove beyond question that molecules really did exist. As a happy byproduct, he would extract a considerable amount of new data about their size, shape, orientation in space, and chemical behavior. Beginning with forays into what is now known as surface chemistry, he began by studying grease films resting on the surfaces of liquids. Using only a tray of water and a dynamometer (an instrument for measuring mechanical force), he floated a light rod on the grease surface. When the oil had formed a film, he maneuvered the rod so that it began squeezing that film. The dynamometer then showed how much electrical force had been needed to do this.

Next, he noted that oil on top of the water behaved in two different ways: it either stayed intact, or oozed over the entire surface of the water, thinning as it spread. Langmuir theorized that the film would keep thinning and spreading until its thickness was equal to the size of a single molecule. At that point, some of the intermolecular forces would prevent any further spreading and thinning. The dynamometer confirmed that the resistance began to increase quickly and very dramatically as the film kept thinning. He realized that the pressure at this thinnest "turning point" was identical for every organic acid. The length of the chains of the molecules of various types of acids was irrelevant. He and his co-workers thereby gained important new insights into the structure of organic compounds such as wax, organic acids, proteins, and so forth. For one thing, Langmuir learned that instead of being rigid and inflexible, the way they could be compressed showed them to be flexible chains, much like ordinary metal winter-weather tire chains.

Why did molecules behave in this way? After much trial and error, he realized that if he replaced the carbon at the end of the chain with an arrangement of atoms characteristic of *acids*, as used in the earlier trials, then and only then would a film form. As Langmuir explained it, "If you have . . . for example, petrolatum or Mujol, it forms little globules on the surface of the water . . . when you compress the film . . . the chains must be vertical. The molecules will then occupy their smallest area." The reason for this is that an acid has an affinity for water, and will thus tend to spread out in order to be in contact with as much of the water as possible. But when the acid group at the ends of the molecule was replaced by ordinary carbon and hydrogen groups characteristic of simple organic compounds, the molecules "lost interest" in the water and remained in one spot as a compact globule.

With these results of Langmuir, the world had one of the earliest genuine insights into the nature of the molecular world. Though much information remained hidden, both Langmuir and other scientists would build on these humble beginnings to work out the molecular architecture of the gigantic molecules that comprise proteins and the nucleic acids.

It has to be pointed out here that Langmuir's accomplishments were all the more noteworthy, since the understanding of organic chemistry and, in particular, the distance between carbon atoms and the atoms to which they are bonded, as well as the structure of covalent compounds (compounds where the individual atoms are bound together by the sharing of electrons) generally, was far less advanced at this time than it would be later on. It would not be until the mid-1930s, for example, that misconceptions about bonding in carbon would be cleared up. Ultimately, these new insights would lead to marvels like the discovery of the structure of DNA by Watson and Crick in the 1950s.

GROWING PRESTIGE AND FUTURE IMPLICATIONS

As these discoveries were taking place, more and more of the scientific elite began to notice him. Langmuir was one of only two Americans (the other being Arthur Compton) invited to attend both the prestigious Volta Congress in Como and the Solvay Congress in Brussels in the fall of 1927. These meetings, intended to bring together the world's elite to discuss the latest research in science, were a sign that the prestige of Americans abroad was continuing the upward climb started by Michelson, Gibbs, and others.[11]

MULTIPLE LINES OF RESEARCH

Encouraged by his success and with no fears that he was spreading himself too thin, Langmuir soon plunged into four different research projects simultaneously. Satisfied with his progress in oil film research, Langmuir went on, in 1919, to study vacuum tubes, chemical reactions at low pressures, the chemistry of surfaces, and the electric discharges in gases.

Of course Langmuir was not the first to study such fundamental issues. American chemist G. N. Lewis, though primarily a chemist, was also active in theoretical physics in the United States (along with J. Willard Gibbs of course) and did some of the earliest

U.S. work in relativity theory. He, along with his student Richard Chace Tolman, completed this work on relativity by 1908.

By 1916, Lewis had analyzed the structure of atoms by suggesting that a central nucleus was surrounded by "cubical" electron shells. Over the years, Langmuir extended Lewis's work at the General Electric laboratories. He theorized that the electron shells were concentric, hoping to have a model that would explain chemical reactions. Using a piece of lab apparatus that he'd designed himself, Langmuir demonstrated that hydrogen has a strong tendency to fill its K shell (the first level or circle of electrons outside the nucleus). Using an elaborate apparatus and ingenious technique, he managed to slit a molecule of hydrogen into two hydrogen atoms. He then showed that the atoms could again recombine and, in so doing, release the energy they'd previously acquired in splitting up. That energy, in turn, generated temperatures of over 3000°C.

When two hydrogen atoms combine or join to form a hydrogen molecule, the electrons of each atom are shared to fill the K shells around each hydrogen atom. And the fact that such enormous energies were required to split the two atoms apart shows clearly how much the hydrogen atoms "want" to remain in that combined state.

Langmuir moved through the periodic table, demonstrating similar tendencies on the part of many other atoms. Lithium, for example, has three electrons. The K shell would fill up with two of them, while the third would start another shell, the "L," and so forth with the larger elements.

This showed that the tendency to fill shells was not unique to hydrogen, but was in fact a universal phenomenon for any element whatsoever. Although the elements in nature of course differ in many ways from each other, they could at least be counted on to follow similar general patterns or "rules" of behavior so far as chemical bonding was concerned. The fact that such regularity existed in nature greatly simplified the study of chemistry, for whenever a scientist finds such a regularity in nature, he or she counts it a blessing. (Interestingly, in further research some years later, this work was one of the few times in his career that he applied quantum mechanics to electronic shell structure. Although it was only partially successful, it shows that Langmuir was not utterly in the dark about the "new" physics.)[12]

Later, the Lewis–Langmuir concept, or the idea that electron sharing is essential in the chemical bonding of atoms to form molecules, had to be radically adjusted and refined to account for anomalies in the behavior of certain elements. However, it was a noble beginning.

THE DOROTHY WRINCH EPISODE

By the late 1930s, Langmuir became embroiled in a most unpleasant episode in the history of science. Though it is little known today, it does say something about how science works—or doesn't work. For a number of years a battle had been raging between Linus Pauling and Dorothy Wrinch over the nature of protein structure. Pauling believed that proteins were polypeptide chains, while Wrinch believed that they had a "cage"-like structure. Her theory became known as the "cyclol" model.

Although Pauling's conception finally won out, during the 1930s it was not at all clear who was right. Langmuir, along with Richard Tolman, Niels Bohr, and other eminent scientists, in fact sided with Wrinch. For a while, the controversy was restrained and polite. In a letter of September 28, 1939, Maurice Huggins, an old colleague of Pauling's, wrote to Langmuir, who was by then associate director at GE, saying, ". . . In spite of my criticisms, I want you to realize that I do consider the agreement Dr. Wrinch and you have obtained between the theory and the Patterson–Marker maps excellent evidence, at lease [sic] for a cage structure of practically the same dimensions as the one she has deduced."

Soon, however, things took an ugly turn. An example of the bad feelings created with this controversy is a cryptic undated letter sent to Wrinch (though probably in March or April of 1956, as Wrinch replied on April 30) and on file at the Sophia Smith archives at Smith College. It is not clear who the letter is from, as it is signed only "Bill," but it clearly shows the type of political difficulties that the Rockefeller selection process for Fellows had created during the Pauling–Wrinch episodes:

> Dear Dorothy, . . . I am still betting on your hypothesis. . . . I believe I have had an experience that must be similar to the one you had when LP [Linus Pauling] interviewed you for the Rockefeller Foundation. I am sure you can appreciate what it means to be in the same boat.—Namely, the experience of having an idea that seems sound, trying to put it across, having it knocked down by a different approach that seems to dispose of it, and discovering *later* that it was *not* disposed of, but that it had not really been communicated, because of differences of approach and particularly because the other party was so sure of himself that he really did not listen well.

On May 1, 1939, Langmuir himself wrote rather testily to Pauling about Wrinch's ideas: ". . . Your summary of the X-ray evidence gives an unfair picture of the situation. You do not discuss the real nature of the assumption that Dr. Wrinch and I made. . . . Although I believe these arguments will carry little weight with you, I think at least you should modify the paper to avoid unnecessary controversy based on misstatements of facts."

Wrinch, of course, did not sit on the sidelines. The Smith archives also contain an (undated) letter from Wrinch to Arthur Lamb, editor of the *Journal of the American Chemical Society*:

> Dear Dr. Lamb, After months of study . . . I have become increasingly aware of the threadbare nature of the attack made by Pauling and Niemann on my work, in your Journal. . . . I wish to protest with great seriousness at the publication of this attack. . . . I will recall to you that the article was sent to Dr. Langmuir in April 1939, but not as a referee and that I saw it only by the courtesy of Dr. Langmuir. I did not make a protest about the contents of this article to you then, since . . . I did not for a moment believe that you would allow publication in view of Dr. Langmuir's letters about it . . . still less did I believe that the tone . . . would prove acceptable to you. I take great exception to this tone. . . . Pauling and Niemann were apparently given a perfectly free hand to attack me last July . . . I ask you earnestly to give me a prompt publication and thereby declare the unfortunate incident closed.

Although her article was published, the whole matter really is a sad one in the history of science. Wrinch not only lost the battle, but managed to alienate just about everyone, including formerly close friends. In a poignant letter to me of May 14, 1986, the great biochemist Professor Dorothy Hodgkin of Oxford University wrote as follows:

> Dear Prof. Serafini
>
> . . . He [Pauling] concluded the structure [cyclol] was untenable from chemical considerations and that she had no answer to his criticisms. . . . I knew Dorothy Wrinch well but I agreed with Pauling in his view of the cyclol theory. Because Dorothy Wrinch and I had been close friends in the early stage of the protein work—1934, 35—I said rather less than others in writing, but our friendly relations were somewhat strained until the structure of myoglobin was established. . . .

It is consummately difficult for a writer to present any kind of accurate view of past controversies as heated as this one. But two things seem clear: that Dorothy Wrinch never really got a fair hearing for her theory, and that Langmuir proved to be a scientist of great courage in defending Wrinch against a huge segment of the scientific community.

He was honored by being appointed to the prestigious post of president of the American Chemical Society. As befitting Langmuir's genius, the scientific world honored him in 1932 with the Nobel Prize "for . . . discoveries and researches within the realm of surface chemistry." He was the first American scientist to capture the laurels who did not come from a university faculty. He died quietly in his home in 1957.

CHAPTER 8

Linus Pauling and American Quantum Chemistry

In 1911, Ernest Rutherford conducted one of the greatest experiments in the history of science at Cambridge University. Not only did it advance understanding in physics, but it provided a model of the atom that led directly to important breakthroughs in the scientific community's grasp of quantum mechanics. The theory of quantum mechanics emerged in 1925 for predicting energy states of atoms and chemical bonding. The theory Rutherford was testing viewed atoms as spheres in which electrons were embedded like raisins in a pudding. The sphere was positively charged; the electrons were negatively charged. Each element was characterized by a different number of electrons. For example, hydrogen had one, helium had two, and lithium had three. The experiment was elegant and simple, involving bombarding a metal foil with the alpha particles from a piece of radium.

In 1909, British physicist J. J. Thomson of the Cavendish Laboratory had already suggested the model Rutherford was testing. Yet Thomson's model was only partially successful. He had been able to build his "raisin pudding" model of the atom to fit the periodic table of the elements, and it did succeed in explaining many of the physical properties of the atom. Yet it had some striking limitations: the Thomson conception could explain neither the nature of chemical reactions nor radiation.

Rutherford, already suspicious of the model, believed that if it were correct, the alpha particles would easily pass through the atoms that composed the foil, with the exception of some that would be absorbed. The reason for this is that according to the Thomson model, the atom was not really solid at all. To the surprise of the scientific community, Rutherford showed that some of the alpha particles recoiled dramatically upon impact, much as if the alpha particles had struck some kind of solid object in the foil.

This discovery led to Rutherford's most brilliant conception—the "compact nucleus." He reasoned that the positive charge of an atom, rather than existing in a diffuse state throughout the atom as Thomson had believed, was massively concentrated and probably restricted to the center of the atom. He calculated that this "nucleus" was probably one-trillionth of a centimeter in diameter.

This model, though better than the Thomson model, still proved inadequate to explain the mechanism of a chemical reaction, since there was as yet no adequate account of how electrons—which might explain chemical bonding—traveled within the atom. Niels Bohr then further improved the model by assuming that the electrons traveled in fixed orbits around the positively charged nucleus, much like planets around the sun. That, at least, offered a theoretical explanation of radiation (the release of energy in the form of particles or waves): the electrons could "jump" from one orbit to another, which, in turn, would account for such radiation. In dropping from a higher to a lower orbit and, by definition to a less energetic state, the atom would give off energy or radiation.

But while this improved Bohr model seemed to answer key questions for the physicist, chemists like Noyes of Caltech, Lewis of the University of California, and Langmuir of the General Electric Laboratories were aware, in 1919, that it still could not shed any light on chemical reactions. Chemists knew that a molecule consisted of atoms in chemical combination. But why did some atoms tend to form only certain combinations? Among other possibilities, carbon could combine with two chlorine and two fluorine atoms to form a fluorocarbon commonly known as "Freon," while carbon could combine only with two oxygens to form carbon dioxide. Nitrogen, on the other hand, could combine with three oxygens to form nitrates. Although the combining power or "valence" for most atoms was known merely through chemical observation, there was, as yet, still no adequate theory to explain *why* atoms combined in the way they did.

Valence is actually a measure of the binding power of an atom, depending on the "vacancies" in the electron orbits or shells around the nucleus. Since carbon's outermost shell, for example, has only four electrons, but has a maximum capacity for holding as many as *eight* electrons, its valence is four.

Many oddities surrounded the phenomena of chemical reactions. For example, some substances such as sulfur and iron could actually have several valences, depending on the conditions. When helium and neon were discovered, it was found that they had no valence at all—they could not combine chemically with any other elements. For that reason they were part of a group of substances then called "inert" gases. (Now they are called "noble" gases, since scientists later showed that under extreme temperature and pressure conditions, the "inert" gases could be made to combine.)

University of California physical chemist G. N. Lewis took an early step in solving these problems when he suggested in 1916, as noted earlier, in the chapter on Langmuir, that the correct model of the atom was one of concentric cubes with electrons stationary at the intersections of the sides. Soon, he developed the idea of "pair" bonding, a useful idea in later studies of chemical bonding in reactions.

Building on this work, Langmuir, in 1919, improved everything considerably. In that year, Langmuir modified Lewis's conception into the Lewis–Langmuir theory. Perhaps the most important change was the idea that electrons were in concentric shells, rather than in cubes. Also, in his view, electrons could move in various ways, around the nucleus. For in Langmuir's atom, as in that of Bohr and Rutherford, the nucleus was in the center, surrounded by concentric shells. Each shell could contain no more than a certain number of electrons. The shell closest to the nucleus could accommodate no more than two electrons. Any atom with an outer shell that was not filled to capacity would seek electrons to fill that shell. In general, Langmuir's views helped make sense of the physical properties of substances such as boiling points, melting points, and so forth better than any idea that had come along before.

Then came Linus Pauling of the California Institute of Technology. Pauling was thoroughly familiar with Langmuir's work and wanted to further understand the nature of the chemical bond.

Among Pauling's most important contributions to the extension or fusion of chemical concepts with physics was, without question, his application of the principles of quantum mechanics to the chemical bond. Pauling was not the first to explain bonding via quantum mechanics. The great paper of Heitler and London had shown, in 1927, that the binding energy of hydrogen–hydrogen bonds can be explained by quantum mechanics. But by 1931, Pauling had extended their work when he proposed that "resonance" between two possible structures is what keeps the benzene ring stable.

By the time Pauling attacked such problems, the understanding of quantum mechanics was still in its infancy and still bubbling with controversy and misunderstanding. Even so, there is an oddity here in that Pauling stuck doggedly to his "valence bond" idea, even when, by 1967, molecular structure was routinely discussed in terms of Mulliken's "molecular orbital" idea. Why? In part, this may be due to classic Pauling stubbornness. When Professor Jack Roberts of Caltech tried to get Pauling to change something in *The Nature of the Chemical Bond*, Pauling replied, according to Roberts, "You know, I don't like to change anything in *The Nature of the Chemical Bond*." A more charitable suggestion is that both were equally useful approaches to bonding. As Katherine Sopka points out, "He [Van Vleck] wrote a series of articles . . . in which he utilized both approaches and another article . . . in which he showed that they were, in fact, intimately related and hence could be expected to give the same formal valence rules."[1]

Many were more absorbed in the philosophical aspects of quantum mechanics—the relation to determinism, for example. Quantum mechanics' "uncertainty principle" was supposed to vindicate free will. Yet mankind was so deeply immersed in determinism that the new science of quantum mechanics led even the great Einstein to cry out that "God does not play dice with the universe."

There are, incidentally, other philosophical paradoxes connected with quantum mechanics. Physicists still debate the oddity of how two electrons, having lost contact with one another, can affect one another in such a way that once one electron is determined to have an "up" spin, the other is caused to have a "down" spin, which is necessary to conserve angular momentum (the momentum of an object as it travels in a circle around a central axis, such momentum increasing with the object's velocity). This, of course, is the ancient "action at a distance" problem in a new guise, usually called the Einstein–Podolsky–Rosen paradox: if you're measuring electron a, how does electron b 50 yards away "know" the effect on a?

In physics, men like Heisenberg and Pauli shattered conventional thinking further with their radical idea that motion might not be "continuous." According to the new science of quantum mechanics that Heisenberg and Pauli helped develop, light might be emitted in discrete bundles or "quanta" of energy, rather than in continuous streams.

And the development of the new science at precisely the time Pauling was passing into the ranks of the world's great scientists is some testimony to the role of luck in Pauling's career. Nor did he fail to notice the profound implications the new quantum mechanics might have for the problems he was interested in. Had Pauling come to Caltech and into the ranks of those working in these areas 20 years earlier, he might never have performed some of his greatest work.

Pauling approached quantum mechanics from still another angle. While physicists were looking to it as a general tool for describing physical events on an atomic scale, Pauling saw even deeper than Langmuir in appreciating the tremendous power it might have in the more restricted domain of the chemical bond.

Aside from Langmuir, the field was bristling with men of genius. If a gulf had existed before between physics and chemistry, many were trying hard to bridge that gulf. Nonetheless, it is true that Pauling made many contributions to this collective effort, and his correspondence shows, over and over again, numerous very specific ideas and suggestions that he made in this area. In a letter of January 28, 1938, to W. A. Noyes of the University of Illinois, for example, he suggests the quantum mechanical idea that there is some uncertainty as to where, in a molecule, an electron is at a given time.[2]

Later in life, Pauling described his own mind during the late 1920s, when he was still in the early years of his tenure at Caltech and just getting into problems about chemical bonding:

> One of the great chemical problems, however—the nature of the chemical bond—remained a puzzling one throughout this period. . . . As I recall my early life, I recognize that I was often satisfied with a very incomplete sort of understanding of the various aspects of nature. . . .[3]

From about 1925 on until the mid-1930s, the problem of the nature of the chemical bond virtually dominated Pauling's scientific interests.

THE SOMMERFELD PERIOD

Still one more key player in the quest for the chemical bond was Arnold Sommerfeld. Pauling's long association with him began early—while Pauling was just finishing his doctoral work at the California Institute of Technology. With that behind him, he wasted no time in heading for Munich and Sommerfeld's cramped but parsimonious laboratory at the University of Munich. Sommerfeld also wasted no time. Immediately, he presented Pauling with a theoretical problem in quantum chemistry. It was a rough one, with Pauling spending many torturous evenings tossing crumpled-up notebook paper on the floor. As Pauling later reminisced:

> Sommerfeld was accustomed of course to providing problems to young people that came, and especially, I guess, Americans; I guess the calibre of some of them there wasn't very high . . . he said [Alexandroff] wrote a paper about the spinning electron around 1900; this is standard old electromagnetic theory. With spinning electrons of different kinds he got different values of the g-factor.[4]

The problem Sommerfeld gave to Pauling was only partially solved by him (though, with the legendary physicist Werner Heisenberg also working on it independently and elsewhere, the problem finally yielded its secrets). But Sommerfeld wasn't just trying to rattle the young American and show him who was boss. It really was a critical problem, piercing to the very core of some of the most fundamental problems in chemical bonding.

Soon, the new quantum mechanics, with its view of electrons as "waves" as well as particles, became a juggernaut carrying Pauling along with it. After 1925, advances in understanding the nature of quantum mechanics as well as its relationship to the chemical bond followed quickly. With the new version of quantum theory scientists, Pauling among them, probed ever more deeply into the mysterious architecture of the molecular universe. This does not refer to the *whole* universe; beyond terrestrial matter, much of the matter in space is nonmolecular, as in the case of black holes, neutron stars, and so forth.

Pauling was able to show, for example, that electron "waves" in resonance with one another could be used to explain complex chemical reactions among organic molecules. In the benzene ring, for example, Pauling showed that the bonds connecting the carbon atoms in the ring were neither single nor double bonds, but resonance "hybrids" of each type. That is, the bonds would have some of the features of single bonds and some of the features of double bonds, just as a mule has some of the features of a horse and some of a donkey. In other words, the bond would resonate between C—C and C=C structures, with the single line between carbon atoms (C) representing a single bond, while the double line represents a double bond. (Interestingly, this work on organic compounds had its first important commercial application in 1930, when industrial chemist Thomas Midgley first produced difluorodichloro methane, or "Freon"—a refrigerant in common use today.)

PROBLEMS WITH BRAGG

However, personal and professional friction with the great British X-ray crystallographer W. L. Bragg marred Pauling's career at this early point. In March 1927, an exchange took place between the two men. It is important to note also that Pauling had, just two months earlier, published his most important paper to date on chemical bonding. In large measure, the paper was a culmination of much of his recent work in applying quantum mechanics to chemical bonding. In the paper, he clearly showed that his work was progressing. As he explained:

> Shortly after reaching Munich I read a paper by Gregor Wentzel in the Zeitschrift fur Physik on a quantum mechanical calculation of the values . . . for electrons in complex atoms. . . . Wentzel reported poor agreement between the calculated and experimental values, but I found that his calculation was incomplete and that when it was carried out correctly, it led to values . . . in good agreement with the experimental values.[5]

In short, Pauling had convincingly demonstrated that the structure of the electrons in even the most complicated atomic structures could be described with quantum mechanics. He would go on to give ever more brilliant descriptions of a dazzling variety of atomic structures and crystals.

However, in his reply to Pauling, Bragg clearly indicated his displeasure with Pauling for failing to give him proper credit for his own research, research that Pauling's paper had relied on. That is almost certainly the source of Bragg's celebrated indifference to Pauling when he visited Bragg's laboratories a few years later.

Pauling charged ahead and by 1928, he was ready to formulate a set of principles that, according to his theoretical research, determined the structure of complicated crystals (such as the silicate minerals tourists see in any rock and mineral shop window). Of these principles, probably the most significant was the one that suggested that the strength of the chemical bond between a positively charged ion (an ion is merely an atom carrying an electrical charge) such as the sodium ion in ordinary table salt, and a negatively charged one (such as the chloride ion in table salt) was a direct function of the strength of the charge on the negative ion. In other words, the stronger the negative charge, the stronger the bond between the two atoms.

The principle, in turn, helped enormously in making theoretical predictions about the properties of ionic substances. (An "ionic" substance is simply anything with a number of ionic chemical bonds—an ionic bond forms when electrons are transferred between two atoms. The "covalent" bond, by contrast, exists between atoms where the positive and negative charge difference is less clear or less sharp and where electrons are shared, rather than completely transferred.) All of this accumulating knowledge on quantum mechanics would soon lead to another great accomplishment of Pauling's—the co-authorship, with Samuel Goudsmit, of the classic work, *The Structure of Line Spectra*. (Pauling actually wrote the entire book; the sections on spectra were based on Goudsmit's work, while the sections on quantum mechanics were based on Pauling's work.)[6]

Pauling continued working for the remainder of his European stay, gathering more information on bond angles, crystal structure, and quantum mechanics. The visit with Sommerfeld in Munich was really the high point of Pauling's stay. After the year was over, he did spend a few weeks at Bohr's institute, and about five months at the University of Zurich, attending the lectures of

Schrödinger, one of the great architects of quantum mechanics, and the distinguished physical chemist Peter Debye, who had already successfully applied quantum theory to the study of the thermal properties of solids. Of course, chemical theory was far from complete in the beginning of the decade of the 1930s. Much work remained to be done in chemical bonding as well as quantum mechanics generally, but the most basic steps in applying quantum mechanics to the chemical bond were now history. After this work, progress in chemistry would be rapid. With this new understanding of how atoms joined to form molecules, the pharmaceutical industry blossomed, as did the new industries of synthetic fibers, and plastics.

THE TOLMAN ERA IN PAULING'S CAREER

Only around 1935 would Pauling allow his obsession with the chemical bond to be eclipsed by his growing fascination with molecular medicine—the idea that many diseases can be understood as due to defects in the structure of certain key molecules.

Richard Chace Tolman, another of the great physical chemists at Caltech (American physicist Richard Feynman would later hold the Richard C. Tolman chair at Caltech), was a key figure here and in Pauling's career generally. Pauling would follow Tolman about the corridors of Caltech, pacing anxiously and introspectively outside his office while Tolman was discussing something or other with an eager graduate student. He even went so far as to imitate Tolman's mannerisms, his casual dress and his halting, staccato-like cadence in the lecture hall.

Tolman had been born in 1881 in Newton, Massachusetts, receiving a bachelor's degree in chemical engineering in 1903 from MIT. He also received his Ph.D. from MIT in physical chemistry from A. A. Noyes, in 1910. Though he had a successful career at Michigan and Illinois, his greatest and most productive period began when physicist Robert A. Millikan, president of the California Institute of Technology, invited him to become professor of theoretical chemistry at the fledgling university. Later, Tolman would become dean of the graduate school.

Tolman's influence on Pauling eventually led to what is perhaps the most famous example of a problem in molecular medicine—Pauling's work on sickle-cell anemia, where he showed that the disease resulted from a defective amino acid (the basic "building block" of any protein molecule) site in the molecule.

And it was Tolman who unwittingly encouraged one of Pauling's most significant and earliest intellectual and psychological conversions. Many years later, Pauling recalled the event in an article published in *Daedalus* titled "Fifty Years of Progress in Structural Chemistry and Molecular Biology":

> There were so many gaps in my understanding the . . . problems mentioned above did not stand out among the hundreds that I failed to understand; often I did not know whether to attribute the failure to myself or to the existing state of development of science. . . . One episode impressed itself on my memory so strongly that I conclude that it had a significant impact on my development. In the spring of 1923 Tolman asked me a question during a seminar. My answer was 'I don't know; I haven't taken a course in the subject.' At the end of the seminar, Dr. Richard M. Bozorth, who had received his Ph.D. degree the year before, took me to one side and said, 'Linus, you shouldn't have answered Professor Tolman the way you did; you are a graduate student now, and you are supposed to know everything.'

A lesson for Pauling: science did not proceed linearly or easily. Descartes, the French philosophical giant, had described this method of approach to knowledge in his *Meditations* and *The Discourse on Method*. One had only to begin by systematically doubting all that could be doubted and then rebuild the edifice of human knowledge on absolutely certain foundations. So reconstructed, it was possible, at least in principle, to know with absolute certainty all that can be known. That had been the way of Newton, Hooke, Darwin, and Einstein. And it was the way of Tolman. Tolman really thought it was possible to know everything there was to know, at least within a particular domain of problems. Perhaps the most dramatic example of the worth of Pauling's studies in molecular biology was the discovery of DNA by Watson and Crick. By their own admission, they merely imitated Pauling's methods to arrive at the solution.

None of this is over: as of this writing, Pauling is still actively pursuing research in chemical bonding, molecular biology, and even nuclear physics. Some parts of his work are, of course, more controversial than other parts. His views on the efficacy of vitamin C on cancer have yet to find wide acceptance. Yet the work on chemical bonding, crystals, and the application of quantum mechanics to chemistry stand as ample testimony to one of the towering figures in American science.

CHAPTER 9

Arthur Compton and "Compton Scattering"

At about the time Linus Pauling was just beginning to think about the nature of the chemical bond, another scientist, Arthur Compton, was clearing away more mysteries surrounding the new science of quantum mechanics. In his classic paper, "A Quantum Theory of the Scattering of X-rays by Light Elements,"[1] Compton points out that ". . . J. J. Thomson's classical theory of the scattering of X-rays, though supported by the early experiments of Barkla and others, has been found incapable of explaining many of the more recent experiments." (The experiments were those Compton himself had carried out. According to his findings, when X rays impinged on the surface of a crystal, the crystal reflected them at a lower frequency than they had when they first hit the crystal. This would be like visible light of one color hitting a mirror, and then being reflected as a different color.)

Compton is referring here to the classical wave theory of light. Since Thomson's classical wave theory successfully explained every phenomenon of light except the photoelectric effect, it is hardly surprising that scientists resisted abandoning it. Still, the photoelectric effect, where light hitting a metal surface causes electrons to be ejected from the surface of the metal, was explainable only on the assumption that light consisted of minuscule radiation "packets." The "wave" concept was useless here. In other words, the photoelectric effect required that scientists look at light as particles rather than waves.

In 1923, scientists encountered yet another oddity that, again, resisted explication by the "wave" concept. As Compton soon showed, it too turned out to be explicable only when one looked at light as consisting of "particles."

If it was a time of radical change in science, it was a time of no less radical change in American industry. While scientists were puzzling over apparent contradictions in classical theory, another, seemingly unrelated development was growing in the United States that would soon affect scientific research in a critically dramatic way. Hardy speculators were rising to riches with the newly discovered oil fields in places like northwest Pennsylvania. Standard Oil, under John D. Rockefeller, was obliterating his competitors with his now-infamous "rebate" arrangements with the railroads, whereby Rockefeller's Standard Oil Company would give money back to the railroads if they promised to ship Standard's competitors at much higher prices than they charged Standard Oil.

Still, Rockefeller would soon prove to be a patron saint of science. It was Rockefeller's money that accelerated progress in the most important areas of science: Rockefeller money all but invented, for example, the field of molecular biology in the 1930s. It was his foundation that gave money to stars like Pauling, Wrinch, Weaver, Corey, and Compton himself, to ferret out the great mysteries of science.

From the viewpoint of scientific progress, the years after the turn of the century were exciting and promising. The quantum approach to physical reality was still new and not fully tested or accepted. Much the same was true of Einstein's concept of general relativity.

During Compton's early and middle years, all of this would change: relativity—at least the special theory—would come to be firmly accepted by the scientific community. Quantum mechanics would soon become entrenched in the scientific repertoire, and the nature of chemical bonding would become clearer with the work of Langmuir, Lewis, Noyes, Pauling, and others. In the 1930s, scientists like Wrinch, Corey, Jordan, and their colleagues would come to better understand the nature and structure of protein and other organic molecules.

EARLY LIFE

Compton was born in Wooster, Ohio, on September 10, 1892. In that year, A. A. Michelson, the American physicist already renowned for his measurements of the velocity of light, was invited to join the faculty of the fledgling University of Chicago. Linus Pauling and Enrico Fermi had not yet been born and Robert Millikan was just entering graduate school. Compton's talent was evident early on and his parents made certain that he had ample tools, books, and instruments to carry on his childhood experiments.

When he was 10, he wrote an essay on the evolution of elephants and by his teenage years he was designing and building model airplanes and devising his own aeronautical theories. Soon, an uncle presented him with a telescope and there began his love affair with observational astronomy, later to culminate in his great work on cosmic rays.

Compton entered Wooster College, principally because his father was dean of the college as well as professor of philosophy. Not surprisingly, he got top grades, achieving distinction in astronomy, philosophy, and theology. In fact, Compton throughout his life was as much interested in philosophy as he was in theology. During his years at Wooster he took all of the philosophy courses offered, even receiving the philosophy prize for his work.

Never the stereotypical "bookworm," Compton actively participated in most college sports. He graduated from Wooster in 1913, entering Princeton for graduate work the same year. Though he had initially been interested in mechanical engineering, by the time he finished his M.A., he had already decided in favor of physics. He progressed rapidly, taking his doctorate at Princeton in 1916. He went first to the University of Minnesota, moving from there to the Westinghouse Laboratories. There he came under the influence of C. E. Skinner, who was at the time trying to organize a corporate-based research lab. Despite the success of the General Electric Laboratories, industry-based but university-like labs were still in their early stages. At Westinghouse, in any case, the company found the idea of such a lab and its practical results not worth the gamble. It was only during the late 1920s and early 1930s, with the rapid spread of radio, that the industrial research lab really mushroomed. In the case of Westinghouse, it was S. M. Kintner who was finally able to mobilize a staff and the resources that allowed the Westinghouse lab to ferret out a wealth of scientific information.

Compton had, however, quit Westinghouse at the urging of his wife Elizabeth before all that came to pass. After a year at Cambridge University, he joined the physics department at Washington University in St. Louis, Missouri. Interestingly, among his personal papers at Washington University are notes he took while making the sea journey from England to take up his new post at the university, which give some idea of the work he intended to do there. The notes show that he was interested in the photon.[2]

In 1923, Michelson, recognizing Compton's genius, invited him to the Ryerson labs at Chicago. They were in daily contact for years and they remained good friends throughout that time, though they never actually collaborated on a project. It was here that Compton conducted the research that soon garnered him the Nobel Prize for physics.

Interestingly, although Compton did spend time at Cambridge University, he did not make the pilgrimage to Europe, considered *de rigeur* at this time, as part of his graduate or postgraduate training. Unlike so many of the scientists of his era, such as Pauling, T. W. Richards, Langmuir, and others, he did not deem it necessary to sit at the right hand of men like Heisenberg, Pauli, or Sommerfeld in exotic places like Munich or Copenhagen.

(This, incidentally, is a bit ironic, since much later on, Sommerfeld would take special notice of Compton's work. In one of the editions of his classic text *Atombau und Spektrallinen*, he devoted fully ten pages to Compton's work on the importance of the work on X-ray scattering for quantum theory, as well as mentioning it often in lectures.)[3]

Always his own man, this independence manifested itself in other ways. Terribly unfashionable as it was, he was a deeply religious Protestant and made no excuses for it. Indeed, he was one of a few distinguished American scientists who, like Millikan, would address religious groups, or at least address religious themes in his lectures. In one of his last addresses, he spoke to the University of Alabama on January 16, 1962, saying:

> Certain it is that our Christian heritage has given Americans a strong incentive to make of their world the best that is possible. Recognition of the value of men and women and of the importance of service on their behalf has been influential in forming our nation's government, in developing our business enterprise. . . . The gospel of service for one's neighbors was a seed planted by Christianity.[4]

WAR WORK

Despite such faith and the pacifism preached by Christianity, Compton's patriotic allegiance was sufficiently strong that he became immersed in the Manhattan Project, believing that the atom bomb was necessary to end the war quickly and save lives, a subject he discusses in his book *Atomic Quest*. Compton, as chairman of a National Academy Committee, presented their report on November 6, 1941. Scientists, government officials, and others had organized this committee to review the military potentialities of atomic energy and it was this committee that generated the vast uranium effort.

Nevertheless, despite Compton's hardy constitution, working on the Manhattan Project became an incredible strain. The inevitable internal friction, with so many scientific egos in one place, was tremendous. Heated exchanges with Feynman, Bacher, and Bethe especially over this or that problem in physics were common. Beyond that, Compton's deep religious sense kept intruding. As his "personal memoir" of June 10, 1945 shows (reprinted in the book *The Cosmos of Arthur Holly Compton*, Marjorie Johnston, editor, Alfred A. Knopf), he had, early on, begun to concern himself with the ethical aspects of the development of the atomic bomb.

Eventually he became chancellor of Washington University in St. Louis, where he had performed his famous experiments measuring the Compton effect—the phenomenon whereby a photon colliding with any charged particle transfers some of the energy of the photon to the particle—nearly a quarter of a century earlier.

EARLY X-RAY STUDIES

The classical theory that Compton helped overturn had, as noted earlier, been almost universally successful in explaining the general features of atomic spectra, and the wavelengths displayed by any electromagnetic radiation. (A rainbow is a perfect example, where the different wavelengths of visible light appear as different colors.) Moreover, the classical Bohr–Sommerfeld theory was in line with commonsense intuitions about nature. Of all the experiments conducted in the 1920s on line spectra, or the visible distributions of electromagnetic radiation (the spread of colors in a rainbow is again a good example), Compton's were among the most important. Compton's research revealed the shortcomings of the Bohr theory and his experiments helped establish some of the tenets of quantum theory on a firm basis (the uncertainty principle, for example, or the idea that both the velocity and position of subatomic particles cannot be known with certainty). And Compton's work helped establish the direction of the development of atomic theory for years to come.

His major work, then, consisted in studying the reflection of X rays from crystal surfaces with an X-ray spectroscope which could measure wavelengths and frequencies—much like the prism spectroscope for visible light. His apparatus consisted of an X-ray tube enclosed in a shield as the source of photons (the particles constituting light) as well as an angular device to detect the photons deflected by the electrons. In his design, an X-ray "gun" made from the metal molybdenum emitted a beam of X rays with a known frequency. The beam was then allowed to strike electrons. Theoretically, any substance would do. But to make things easier, Compton wisely chose substances that held their outer electrons very weakly, as was the case with atoms like carbon.

The reason this was a wise choice is that in his experiment, when a photon collided with an electron, they behaved like billiard balls colliding with one another, where energy from one ball is transferred to the other. If the nucleus of the atom held or bound the electrons too tightly in a fixed position, this energy transfer between photons and electrons would have been extremely difficult to detect. In Compton's experiment, he directed the primary X-ray beam against a graphite target and studied the rays coming away from the target at right angles to the direction of the primary X-ray beam.

During the X-ray bombardment of the carbon with rays of a known frequency, he found that at some angles the X rays were bouncing off the target with a frequency lower than when they left the X-ray "gun." While this appeared mysterious and bewildering viewed from the perspective of classical electromagnetics, Compton did first try to reconcile the results with the classical theory anyway. He tried by assuming that the electrons have a certain size and that X rays emitted from different parts of the electron "interfere" with each other in such a way as to give the observed results. While some criticized this work, others were quick to voice their approval. X-ray crystallographer W. H. Bragg, for example, in a letter to *Nature* on May 27, 1915, wrote:

> I believe Mr. Compton is right in ascribing the rapid decline in the intensities of the X-ray spectra as we procede to higher orders to the

fact that the atom should not be treated as a point, but as a distribution of electrons in space. . . .[5]

However, the theory was simply too complicated. For one, it violated a fundamental principle of scientific theory—Ockham's "razor," which demands that explanations must not be more complicated and must make no more assumptions than necessary to explain the phenomenon at hand. But that's exactly what Compton had to do in order to make this idea work. He had to postulate an "electron size" or a new constant that was a function of the frequency of the X rays striking the carbon electrons. He soon realized that this was becoming needlessly complex, and he made a monumental leap: he abandoned the classical picture and turned to the quantum theory.

Compton then assumed that the X rays were packets of radiant energy, rather than waves. A light wave, from the classical viewpoint, distributes its energy continuously. Photons, however, do not. Although it alludes to some quaint museum concepts, one of Compton's descriptions bearing on this phenomenon runs this way:

> one hot, calm July day, my brother . . . dove into the deep water. By the time the ripples had spread . . . to where I was swimming a half-mile away, they were too small to notice. You can imagine my surprise when these insignificant ripples . . . suddenly lifted me bodily from the water and set me on the diving board.[6]

He goes on to explain that there is no need for the "ether" (the substance once thought to be spread through all of space in which light waves traveled), believing that the inertia of photons would take them vast distances without any diminution of energy. Indeed, the amount of energy carried in this way could be calculated by a relatively straightforward mathematical equation. Compton also theorized that packets of X rays were smashing into electrons. Using Compton's own analogy, when billiard ball "A" collides with ball "B," the impact shifts some of the energy of the first ball to the second. He then applied the principles of conservation of momentum and energy and made relativistic corrections (necessary because of the great velocities of the particles involved). In this way he showed that the scattered "quantum," or packet of energy, would head away at a specific angle with the X-ray beam. He further showed that the "recoil" electron would travel at an angle demanded by the conservation laws.

By extending his billiard ball collisions analogy, Compton further reasoned that when the X-ray photon smashed into one of the electrons in a carbon atom, the photon had to have lost energy in order to accelerate the electron as it had in fact done. He thus demonstrated that the energy of the photon is reduced by the same amount that the kinetic energy of the recoil electron from the carbon atom is increased—much like what one sees in shooting pool. This diminished energy of the X-ray photon then showed up in the laboratory as a change in frequency of the X ray.

Compton, using the billiards model, successfully tested the theory. Further experiments confirmed his work easily and elegantly. Eventually, Compton's discovery was confirmed by Indian physicist C. V. Raman for visible light.

For this, Compton and C. R. Wilson of England shared the Nobel Prize in 1927. (Wilson's introduction of the cloud chamber, later to become a common laboratory device for detecting charged particles, helped support Compton's theories.)

QUANTUM MECHANICS AND THE COMPTON EFFECT

But the results of Compton's work extended beyond the theory of photon quanta. Not only did his crucial experiments validate his account of X-ray photons, but the quantum theory itself was further supported. For one thing, Compton's work constituted experimental evidence for the Heisenberg "uncertainty principle."

Compton reasoned this way: to locate an electron precisely it had to be bombarded with high-frequency photons. Because of the inverse relationship between frequency and wavelength, such photons will have very short wavelengths. And for a photon to be seen, it must then enter an observing microscope after colliding with the electron. Because of the Compton effect, the electron recoils as described above. But we can't fix the recoil of the electron precisely since we have merely a rough idea of the direction of the scattered photon. (The reason for this is that the photon can enter the viewing part of the observing microscope from many directions, given the large apertures of the typical microscope lens.) Thus, because of the Compton effect, we can precisely fix neither the recoil nor the velocity of the electron if we locate it precisely—exactly as Heisenberg had claimed.

WORK ON MAGNETISM

However, the "Compton effect" was not the only contribution that Compton made to physics. By the early 1930s, the concept of Fermi–Dirac statistics, or the idea that certain kinds of subatomic particles cannot occupy the same orbit or "quantum state" in an atom, with its eventual clarification of many problems in ferromagnetism, appeared. The theory soon proved to be enormously practical. Once the properties of metals became well understood, the field of solid-state physics, with its great breakthroughs in semiconductors, superconductors, superfluidity, etc., came into its own.[7] During the 1930s, for example, many scientists tried to find an adequate theory of ferromagnetism—the type of magnetism displayed by ordinary bar magnets.

John Slater of MIT, for instance, severely chastised Linus Pauling of Caltech over differing approaches to this significant question in solid-state physics. Pauling had, on November 30, 1953, sent Slater his paper "A Theory of Ferromagnetism," hoping for his approval. But Slater had torn it to pieces, saying, among other things, that "When you spoke to me about it in Washington last spring, I indicated that I didn't like it very well, and I'm afraid that I still don't. . . . I think my fundamental objection is the way in which you give the impression that an extremely sketchy calculation must be right because you can get some numbers from it which are in agreement with experiment to three figures."[8]

Arguably, neither had taken sufficient account of Compton's 1921 work, possibly because Compton himself was never one to go to unusual lengths to promote his own work. In fact, Compton's observations contributed substantially to the victory of Slater's, rather than Pauling's, approach to ferromagnetism. By careful investigation of the intensity of X rays diffracted by magnetized and unmagnetized magnetite and silicon steel, Compton and his associates could at least speculate intelligently that the source of the magnetism could be explained only by a leftward tilt of the spin of the electrons, in line with Slater's results. However, it wasn't until Lev Landau's work of 1935 that anything resembling a complete explanation of ferromagnetism had even begun to emerge.

(Whatever the case, as a human interest aside, it may be just these uncertainties about the nature of magnetism coupled with a full awareness of Pauling's abilities generally, that led Slater and Compton to try, unsuccessfully, to attract Pauling to MIT. In a letter of January 21, to Pauling, Slater said ". . . the first thing I suggested [upon returning to MIT after a leave] . . . was that we try to get you . . . Compton at once agreed, with enthusiasm. . . .")

Later, using X rays as a measuring instrument, Compton "explored" the interior of the atom, making extremely precise measurements, for example, of the distances between atoms in a calcite crystal. In doing so, he extended the theoretical understanding of the early experiments on crystal structure begun by Roscoe Dickinson, Tolman, and others early in the 1900s.

COSMIC RAYS

By the late 1920s, Compton was just starting to become interested in cosmic rays. The great value of cosmic ray studies is the information they provide about such things as the importance of Einstein's theory of relativity for electromagnetic theory, the nature of elementary particles, and the nature and structure of the atomic nucleus. Scientists already suspected certain things about cosmic rays. Rudolph Hess, for example, had suggested that they originated from outside the atmosphere, although no one had really worked out their nature in detail. However, others had already made forays into the field: Oppenheimer, Bacher, and, most importantly, Millikan were interested in the nature of cosmic rays.

It proved to be one of the more unpleasant phases of Millikan's career, since Millikan found himself on the losing end of a battle with Compton over the nature and origin of cosmic rays, and Millikan was not a good loser. In fact, it was Millikan who first used the term "cosmic rays," at a meeting of the National Academy of Sciences, held in Wisconsin in 1925. For many years, Millikan had defended the view that cosmic rays were primarily photons with no electrical charge. This was a natural enough assumption. Scientists already appreciated very well that cosmic rays had tremendous penetrating power—far greater than X rays, for example. By treating them as photons, Millikan, along with his associate Cameron, would compute the individual energy of a cosmic ray, since that energy was a direct function of its penetrating power. He could also calculate the energy freed in certain atomic reactions. In fusion reactions, for example, enough energy is freed to yield exactly what is required to give a photon a penetrating power exactly equal to the weakest observed cosmic rays.

Naturally enough, these parallels suggested what Millikan later came to call the "atom-building" hypothesis, which required that cosmic rays be considered photons which arose when matter was converted to energy in accord with Einstein's equation $E=mc^2$. In this way, uncounted billions of helium atoms are supposed to be constantly assembled in a cosmic factory by the fusion process perpetually going on in the vastness of interstellar space—a conception that fit very well with Millikan's religious view of the cosmos. (This approach meant that the universe was becoming more ordered, as if by the hand of a benevolent deity, rather than constantly deteriorating.)

As a theist, Millikan was impressed by the "Argument from Design," propounded in the High Middle Ages by Thomas Aquinas as a proof for the existence of God and defended vigorously in the 19th century by William Paley. In this argument, Aquinas and Paley reasoned that the apparent design, order, and "purposefulness" of the cosmos was clear evidence of the Hand of God at work. Also in this view, the universe tends not to disorder but to order. It could not be otherwise as all of nature, for Aquinas, for Paley—and for Robert Millikan—offered overwhelming evidence for the existence of an Almighty Designer. A "heat death," where the universe finally winds down to a dead, cold state, was, under such a theistic view, an unthinkable heresy. In Millikan's words, a cosmic ray was "the birth cry of an atom"—a "signal broadcasted throughout the heavens" (from his 1928 *Science* article "Available Energy").

On December 29, 1930, Millikan spoke in defense of his views at the Meeting of the American Association for the Advancement of Science. He expressed satisfaction that "the Creator is continually on His job." This announcement, coupled with Millikan's brash proclamation of his theological faith, soon attracted worldwide attention. Though there were various technical problems with the atom building theory, at first it did seem plausible. As Millikan and his associate Cameron put it, "Nucleus-binding is a phenomenon which, in some yet unknown way, is favored by the extreme and thus far unexplored conditions of low temperature and density existing in interstellar space."[9] After the initial buildup of atoms, Millikan reasoned, other well-known building processes could then occur in one vast and unending cycle. After the formation of hydrogen gas, helium would be formed in the unimaginable cold of space. Then, inevitably, and in a parallel way, heavier atoms like iron and oxygen would be formed. These, in turn, would produce even more powerful and more penetrating cosmic rays. Ultimately, gravitation pulled the atoms together into nebulas and stars.

It was a grand picture. But there was a hitch; it contradicted much of the then current scientific thinking. The theory of the time held that the entropy of the universe was gradually increasing, in accord with the second law of thermodynamics. Since entropy was a measure of increasing disorder, a final state would be reached where the universe was completely disordered and chaotic—the infamous "heat death," where all life, all energy-producing processes would come to a complete and final halt. That was Compton's view, and the view of most scientists at the time. But it was not a concept of the universe Millikan liked; God would never let that happen to His Creation.

Naturally, the dispute soon turned to controversies over the evidence supporting either view. The locus of the soon-to-erupt hostilities between Compton and Millikan centered on the so-called "latitude effect"—whether or not the magnetic field of the Earth in any way affected the intensity of cosmic rays of different points on the globe. Compton, against Millikan, did not believe that cosmic rays were neutral photons. Rather, he thought they were charged particles. (Whether this charged particle was an electron or not, would be a question to be answered in future research.)

If Compton was right, the Earth, like a giant magnet, should bend the path of such charged particles in their voyage to us from deep space. But if Millikan was right, and cosmic rays were uncharged particles, no such bending would occur. There was already experimental work bearing on this question, much of it initially favorable to Millikan. Early in the controversy, German physicists Walter Bothe and W. Kolhorster had traveled from Hamburg to Spitsbergen performing their experiments, but without being able to show a latitude effect. At around the same time, another researcher, geophysicist Kennedy, traveled from Australia to the Antarctic—also with no data even remotely suggesting a latitude effect.

Other promising results for Millikan's theory came from research with his co-worker G. H. Cameron. They performed experiments in the Andes and on an ocean voyage to Bolivia, as well as near California's mountain lakes, and found that the rays entered their instruments in equal amounts no matter what direction they came from. They also appeared to be unaffected by latitude or anything in the atmosphere.

So far so good for Millikan. If cosmic rays really were uncharged photons, variations in the strength of the Earth's magnetic field would not swerve them from their course and no latitude effect would show up.

Still unsatisfied with the evidence, Millikan continued his quest for evidence supporting his view. The year was 1932. Like 1925, it was a critically important year in physics. Not only did we have Chadwick and the neutron and Anderson and the positron, but William Fowler was beginning to do important work at Ohio State in engineering physics, and would later receive the Nobel Prize.[10] In that momentous year, Millikan took his own entourage to the Arctic Circle, also sending a graduate student, Victor Neher, to South America in search of a latitude effect. Fortunately, Neher found no such effect. Again Millikan was delighted.

However, Compton had not been idle during this time. As early as 1926, Compton himself had traveled to India, functioning both as lecturer at Punjab University and as a Simon Guggenheim Fellow. It was on this trip that he made some important early, though not conclusive, data accumulations appearing to support Compton's theory, from high in the Himalaya mountains.[11]

In 1928, however, Bothe and Kolhorster devised another experiment to see if cosmic rays did, in fact, consist of charged particles. The result was their famous "coincidence" experiment. By arranging two Geiger counters (hence the term "coincidence," since a cosmic ray would hit both counters) to register the passage of any charged particles, they were able to measure the penetrating power of the radiation and to suggest, on the strength of some real evidence, that cosmic rays actually were charged particles instead of photons, thus casting doubt on Millikan's theory. Later, the great Italian physicist Bruno Rossi supported these ideas. And when Rossi spoke at the Rome conference, he made it clear that he regarded Millikan's view as untenable.[12]

Though this was a setback, other results continued to be, on the whole, favorable for Millikan. The intensity appeared to be unchanged by any variations in location or altitude and, hence, by any variations in the Earth's magnetic field. Finally, he ran a series of experiments at the Hudson Bay, again with no latitude effect.

By this time, the prestige and public interest in the controversy had advanced considerably, doubtless because Millikan and Compton, as Nobel Prize winners, carried enormous prestige. Not only did Millikan have the vast resources of Caltech at his disposal, but the Carnegie Institution was supporting him as well. Millikan's group performed early experiments in the Andes, both at sea level and near the summit. Cosmic-ray electroscope in hand, Millikan further journeyed to the Hudson Bay, near the north magnetic pole, still in search of the latitude effect. He and his team performed yet another experiment in California and still the results were favorable.

In addition to the above-mentioned work, about the only other work that seemed to support Compton's view was that of Dutch scientist Jacob Clay. Clay did his work in 1927 and he claimed to have found variations in the rays between the Netherlands and Java. However, although Clay and his assistant Corlin seemed to have detected the latitude effect Compton's view required, further work in the same geographical area detected no evidence at all of a latitude effect.

In 1930, Compton decided, once and for all, to settle the problem. He radically intensified his cosmic ray work, armed with considerable funding from the Carnegie Foundation. Compton's powerful and poetic description of the cosmic ray question is worth quoting:

> Cosmic rays come from outside the earth, no one knows how many thousands or millions of light-years away. They speak to us in a language that we are only beginning to understand, telling us of the conditions where they were born. Whether this birth occurred within the heart of the atom, or in the vast regions of interstellar space, we have not yet learned.[13]

TRAGEDY MARS THE RESEARCH

By the annual December meeting, in 1932, of the American Physical Society, relations between Millikan and Compton had become so bad that Millikan even refused Compton's offer of a handshake. Still, only more experiments would tell which theory was correct. To make the search for cosmic rays more reliable, Compton designed new detecting apparatus. An essential consideration was the electrical conductivity of air. Compton's team used a steel shell with a two-inch radius. They then filled it with the inert gas argon under extreme pressure. The theoretical key was that a cosmic ray would give the gas a minuscule—but detectable—amount of electrical conductivity. This, in turn, could be measured with an electrometer. In regions of high-intensity bombardment, the current would be strongest, and vice versa.

By February 1932, he was ready. Compton put together a number of research teams, each of which would go to various parts of the globe in search of the elusive "latitude effect." Most famous of these was Admiral Byrd who, along with his assistant, Professor Pouler of the physics department of Iowa Wesleyan, would look for the effect in the Antarctic. Compton himself headed for such far-off lands as Maui, Peru, Panama, New Zealand, northern Canada, and northern parts of the United States.

Soon tragedy struck. A freak accident on a glacier near Mount McKinley took the lives of several members of Compton's team, including John Koven and his assistant Allen Carpé, an engineer at AT&T. Nevertheless, they had collected enough information to devastate Millikan's theory. They found a strong latitude effect near the pole. Also, Ralph Bennett of MIT sought the latitude effect in Alaska and Colorado, while E. O. Wollan of the University of Chicago conducted several precise measurements in Spitsbergen and Switzerland. These studies proved that the magnetic field of the Earth drastically affects the intensity of cosmic rays. As Compton summed it up:

On bringing together the results of these expeditions, it was found that the cosmic-ray intensity near the poles is about fifteen per cent greater than near the equator. Furthermore, the intensity varies with latitude just as predicted due to the effect of the earth's magnetism on incoming electrified particles. At high altitudes the effect of the earth's magnetism is several times as great.[14]

In September 1932, the *New York Times* said, in large bold type, simply "Compton brings . . . data on the cosmic rays, holds . . . Millikan is wrong." As Compton himself said at the time, "My work on the arctic barren lands shows . . . the cosmic ray is an electron, not a wave as Dr. Millikan believes. . . . The difference shown by my experiments will be a severe blow to Dr. Millikan."

Compton cited, as proof of his view of cosmic rays, the work of Bothe and Kolhorster on the latitude effect due to the Earth's magnetic field, as well as the changes in ray intensity with altitude. This, to Compton, proved that rays could not possibly enter the stratosphere as photons. Considering Millikan's public and long-standing hostility to Compton, this blunt statement was understandable on Compton's part. Millikan had never reacted well to criticism.

According to historians of science Russo and De Maria,

Quite unfairly Millikan did not report [in an address in Atlantic City, N.J. and later in the *New York Times*] Clay's measurements, nor did he cite Compton's results. He cited rather the observations of E. Oeser of the University of Gottingen, . . . who reported no change whatever in his electroscope No. 1, that was air tight, but a lowering of 17% in the equatorial regions for a second electroscope which was definitely proven to be leaking and which therefore contained more rarefied air in the hotter areas than in the colder ones.[15]

Interestingly, according to Russo and de Maria, "If one scientist deserves the title of discoverer of the latitude effect, he is Jacob Clay, the director of the physics laboratory at the Technische Hoogeschool in Bandung, Java."

Even as late as 1939, Millikan was still clinging to his hypothesis. The physics community was growing frustrated at Millikan's intransigence. One of these was J. Robert Oppenheimer. In a letter to his brother on June 4, 1939, in an obvious slap against Millikan he wrote:

> Dear Frank, Only a very long letter can make up for my great silence, and for the many sweet things for which I have to thank you. . . . As you undoubtedly know, theoretical physics [stands firm] . . . with the . . . conviction, against all evidence, that cosmic rays are photons . . . is in a hell of a way.[16]

His success notwithstanding, Compton continued to seek absolute verification of his theory. He therefore continued to travel widely, often to locations as disparate as Panama, Peru, Ecuador, New Zealand, and Hawaii. The southern hemisphere was particularly important, since he discovered that cosmic ray intensities are less near the Equator than at the poles.

An earlier blow for Millikan had come in December 1931, when J. T. Carlson and Robert Oppenheimer presented a paper at the Physical Society meeting in New Orleans. In this paper, the pair, for the first time, offered overwhelming evidence for a new view: cosmic rays were not photons at all. Instead, they might be composed of neutrons and electrons, in accord with Compton's view. Eventually, Compton reached a dramatic and critical new stage in his work. With stratospheric data obtained in the summer of 1933, he proved that the variety of particles making up cosmic rays was even greater than he or Millikan had suspected. But this made things even worse for Millikan. None of these "new" cosmic rays were made of photons.

POSTWAR COSMIC RAY WORK

This was, of course, not the end of work on cosmic rays. Work continued on them intermittently throughout this century. It received a shot in the arm in the series of conferences on theoretical physics organized after World War II. At one of these, held in 1947 on Shelter Island, Oppenheimer presented new work on cosmic rays. As he stated at the time:

> It was long ago pointed out by Nordheim that there is an apparent difficulty in reconciling on the basis of usual quantum mechanical formalism the high rate of production of mesons in the upper atmosphere with the small interactions which these mesons subsequently manifest in traversing matter. To date no completely satisfactory understanding of this discrepancy exists, nor is it clear to what extent it indicates a breakdown in the customary formalism of quantum mechanics. It would appear profitable to discuss this and related questions in some detail.[17]

The point of Compton's work then comes to the fore: what his research in the 1930s had proven was that the experiments exploring how cosmic rays interact with the atmosphere could be a valuable source of data on elementary particles.

Indeed, scientists found the "mesotron," a particle intermediate in mass between a proton and an electron, because of cosmic rays studies in 1937. But there were grave problems in identifying them. In that year, Henry Neddermeyer and Anderson said that these new particles, while neither protons nor electrons, were nonetheless charged. The scientific world believed this because of their ability to cut through a 1-centimeter plate of platinum. This caused problems because according to current theory, the platinum should have absorbed the particles.

For a while, the most plausible explanation was that the part of current scientific theory that held that they should have been absorbed, simply did not apply at such high energies. Among those who believed this was British physicist P. M. S. Blackett, saying that the particles were, in fact, electrons. Yet meson observations showed that their mass was about 200 times that of an electron, or about one-tenth of a proton. Also, the particle was highly unstable and had both a positive and a negative form. Additionally, its lifetime was only about 10^{-6} seconds. The "mesotron" (now called the "meson") had already been theorized by Japanese physicist Hideki Yukawa to be the central element in explaining how the nucleus is held together.

But, if the mesons are produced in great quantities by cosmic rays entering the upper atmosphere, why was it that so many managed to penetrate virtually all of the layers of the atmosphere without being absorbed by the atmosphere, especially since the then-current thinking on the meson indicated that it should be so absorbed? Why, also, did it not interact with anything? Things got even worse when Italian physicists showed that carbon plates were capturing negatively charged mu mesons—before they could be captured by the nucleus.

It was at the Shelter Island conference that physicist Robert Marshak, a former graduate student of Hans Bethe, offered his famous "two-meson" theory as a solution to this contradictory evidence. In his view, there were two kinds of mesons, differing slightly in mass. He theorized that it was the heavy meson that was responsible for the nuclear forces and, therefore, the one to which Yukawa was referring. It was, then, the lighter meson that was getting through the atmosphere. Marshak realized that unless Yukawa had gone badly astray—and there was no reason to think he had—the heavier meson had to exist. Yet there was clearly another type of meson getting through the atmosphere which did not fit the requirements of Yukawa's theory. Hence, both experiment and theory said there had to be two kinds of mesons.

Not too many years after this meeting, physicists found the heavier, or pi meson in nature, thus corroborating Yukawa's theory. This climax came at Bristol University in England, via new detection techniques developed there. Finally, with that technique, the pion appeared, as well as Yukawa's nuclear meson which decayed into the meson found in cosmic rays.[18]

The discoveries of the new particles were not altogether happy events in physics. Since the formulation of "Ockham's razor" in the Middle Ages, scientists have been loath to accept any more theoretical entities than necessary. Psychologically, then, physicists found the new particles were making them uncomfortable. As Maurice Goldhaber remembered, at a symposium in Minnesota:

I remember being . . . shocked when it dawned on me [1934] that the neutron, an "elementary particle" as I had by that time learned to

speak of it, might decay by β-emission with a half-life that I could
roughly estimate . . . to be about half an hour or shorter. . . .[19]

And Goldhaber was not the only one. Even Dirac in his original
paper on the electron regarded this new entity as a proton—long
known to exist. He even tried to speculate that, under certain
conditions, a positron could be the size of a proton. It was not until
Herman Weyl's work proved they must have the same mass as
electrons, that science abandoned this idea.

To some extent, of course, this confusion and its attendant
slow progress had political origins. The depression in Europe and
America was severe, fascism was on the rise in Europe, and
governments were forcing scientists to turn more and more to war-
related work.

Though the application of cosmic ray findings to an understanding of nuclear forces went far beyond what Compton had
done, none of the progress at Shelter Island and afterwards would
have been possible without Compton's work.

COMPTON AND THE BIG BANG THEORY

It is essential to realize that the controversy between Millikan
and Compton turned out to have implications far beyond the
question of the origin and nature of cosmic rays alone. The
controversy had powerful implications for the great cosmological
questions about the very origins of the universe itself. Early in the
controversy, another view appeared—a cosmological hypothesis
of great power, known today as the "big bang" theory. The young
Belgian priest Abbé Goerges Lemaitre first suggested this idea.
Although he originally studied theology and became a priest, he
retained a boyhood fascination with astronomy that he'd never
abandoned. After receiving Holy Orders, using a scholarship
given him by the Commission for the Relief of Belgium, he was
able to study physics and astronomy at Harvard. By the early
1920s, Lemaitre had begun studying the work of Russian mathematician A. Friedmann. In the spirit of Millikan, Lemaitre be-

lieved that the principles of science and the principles of theology were just alternate ways of arriving at God's Truth.

Lemaitre's idea was that the universe originated billions of years ago in a cataclysmic explosion where bits of matter where thrown out.[20] The theory received confirmation by Edwin Hubble using the great telescopes at Mt. Wilson, where Hubble found that objects such as galaxies, because of their redshift, appeared to be moving away from one another with tremendous velocities. The cosmic rays, then, are merely part of the matter spewed forth by the "big bang."

Sadly, Compton would not live to see all of these awesome implications of his research on cosmic rays. Though the cosmic ray research hardly ended Compton's theoretical work in science, the period of his greatest work was now over.

On April 10, 1953, he resigned as chancellor of Washington University and began his term as Distinguished Service Professor of Natural Philosophy the next day. In his letter to the board of directors, he said:

> . . . ever since the recent war was drawing to a close, I have been eager to undertake a task of a scholarly nature for which my experience has seemed to give me certain special qualifications. This task has to do with the relations of science to human affairs. . . . Eight years ago, I turned away from the opportunity to work on these problems because I saw in helping to reorganize the work at Washington University an undertaking of such importance and urgency that it could not be refused.[21]

In the 1950s he became involved in politics. Indeed, he was one of the earliest critics of Senator McCarthy, attacking the latter on March 29, 1950, saying of McCarthy's onslaughts,

> The lethal character of such attacks is illustrated by Senator McCarthy's broadcast attack upon our State Department. Nothing could be more effective in destroying the influence of democracy abroad, or the spirit of loyal Americans who so believe in their freedom that they will devote their lives. . . .[22]

And when numerous other scientists got caught in the tangled net of McCarthyism, Compton, often cooperating with the American

Federation of Scientists, helped as best he could. Consider the following comment, in a November 25 letter from chemist Willard Libby, later to win the Nobel Prize for his discovery of radiocarbon dating, to Secretary of Commerce Lewis Strauss (from the files of the Atomic Energy Commission). Libby's comment, referring to such problems, is typical: "We have arrived at the point where I think people pretty well understand what the situation is. It is too bad. Arthur Compton called me yesterday about the resolution the American Federation of Scientists passed yesterday. He was doing his best for us."[23]

And it was doubly odd for Libby to have been in such a position. For Libby was always a staunch patriot, often siding with the Atomic Energy Commission position on the widely publicized 1950's "fallout" controversy, against opponents of such things as nuclear testing. In a letter of May 6, 1955, e.g., Libby says, "All in all, I believe we are justified in saying that although genetic effects are unknown, the test fallout is so small as compared to the natural background . . . that we really cannot say that testing is in any way likely to be dangerous. . . ."

Doubtless, many were drawing courage from Einstein's forthright stands on the excesses of the lunatic right. In a historic letter to Linus Pauling of May 21, 1952 (regarding Pauling's own political problems), Einstein says:

> . . . It is very meritorious of you to fight for the right to travel. The attitude of the government corresponds, of course, to the state of transition toward a kind of totalitarian state in which we find ourselves. . . . Behind this attitude is the intention to crush the Soviet Union by war. . . . The fact that independent minds like you are being rebuked equally by official America and official Russia is significant and to a certain degree also amusing. . . . I am looking forward with pleasure to seeing you again. . . .

In 1954, upon learning of the death of his brother Karl, Arthur Compton said in tribute:

> I have wondered why I have really not found myself sad on Karl's account for having been cut down at an all-too young age. The loss is heavy, especially so to me myself, who relied on him in so many

ways, including the need for . . . an understanding friend. I think the
answer to my question is that of Solon. I truly see Karl's life as a
remarkably happy one, full of the joy of doing his part. . . .[24]

He himself died on March 15, 1962, after an illness of several years.
Through this humanitarian and political work as well as his work
on the "Compton effect" and on cosmic rays, Compton showed the
great depth and breadth of his interests and the range of his
genius. Not surprisingly, the physical community awarded him
numerous honors throughout his life and his place as one of the
great physicists of all time is secure.

CHAPTER 10

Davisson and Germer and Quantum Physics

EARLY LIFE

Along with the work of Pauling, possibly America's greatest contributions to the new science of quantum mechanics came from Bell Lab scientists Leslie Germer and Clinton Davisson. Upon graduating from Cornell with honors in 1917, Germer accepted a research post at Western Electric.

Davisson was born in Bloomington, Illinois, on October 22, 1881, and attended the University of Chicago where he came under the influence of Robert Millikan. He was "delighted to find that physics was the concise, orderly science he had imagined it to be. . . .[1] Soon he began teaching physics and doing research on salts at Princeton, where he was further influenced by the distinguished physicist Owen Richardson and James Jeans. After receiving a Ph.D. in 1911, Davisson joined the faculty of the Carnegie Institute in Pittsburgh, leaving there in 1917 to conduct research at Western Electric, a division of AT&T.

As is often the case, serendipity helped vault them into the ranks of the scientific elite. Davisson had been asked to study

171

electron emissions from the cathodes used in the earliest models of vacuum tubes. With the assistance of Charles Kunsman, a Ph.D. physicist[2] (who eventually left the company and the project in 1923 because of his dissatisfaction with the slow progress of their work), he designed a "gun" constructed to shoot a beam of high-energy electrons at a metal grid. Using nickel as a target, they trained the beam of electrons on several facets of the nickel crystal. The choice of nickel was easily made. As Germer pointed out years later, "Nickel surfaces have been studied more intensively than those of any other metal, not only because nickel is unusually simple but also because it is known to have an 'active' surface and is widely used as a catalyst. . . ."[3]

A LUCKY BREAK

However, although they had been careful in placing the whole apparatus, including the electron "gun" and the nickel target, in an evacuated tube, an accident soon occurred. The tube split, thereby destroying the vacuum. As their February 5, 1925 notebook entry shows, they realized they would have to suspend their work for at least a year to repair and redesign the experiments.[4] Of course, they did eventually repair the apparatus and resume the research.

But something mysterious had happened: the nickel crystal had changed so that incident X rays behaved differently than before. After resuming the experiment, the pair found something they had never expected. In all previous experiments, the X rays bounced back in "specular" fashion, i.e., they behaved much like ordinary rubber balls bouncing off a wall.

This time, however, classical physics could not even come close to explaining what they found. They discovered that not only did the angle of the bounceback vary (as expected), but the *strength* of the "bouncing" varied at different angles of the bombardment as well. This was a great discovery, since this behavior was exactly what one would expect if particles could behave like *waves*—one of the major predictions of the new science of quantum mechanics. As Davisson amusingly summed things up in his article, "Are Electrons Waves?": ". . . it is all rather paradoxical and confusing: we must believe not only that there is a certain sense in which rabbits are cats, but . . . a sense in which cats are rabbits."[5]

Also, as Davisson and Germer noted, by using electrons to bombard nuclei, they were merely following the lead of the distinguished Rutherford. It was Rutherford who had first realized that the nucleus of an atom could be explored by bombarding it with a certain type of particle—in Rutherford's case, the "bullet" was the nucleus of a helium atom. Germer and Davisson used the same idea in their work, merely substituting electrons for Rutherford's helium nuclei. (Certain "bullets" only work with certain targets.) Davisson and Kunsman then plotted a curve that proved that the quantity of electrons scattered by the nickel grid increased immensely at a few specific angles of firing. They subsequently published their results in *Nature* in 1927, *The Physical Review*, and other journals.

X-RAY CRYSTALLOGRAPHY

Their general technique was not new; Max von Laue had conducted seminal studies with X rays about 15 years before the *Nature* article. At the time, von Laue established that an X-ray beam, when striking a crystal, will be scattered by atoms in different planes of the crystal. The same ideas of course governed the Davisson and Kunsman approach, but they now realized that the nickel crystal had been subtly altered so that ten new facets had been formed at the places where the electrons hit. Reasoning as von Laue had, they concluded that the new scattering patterns of the electron intensities must have been caused by the arrangement of the atoms in the nickel crystal, rather than by any intrinsic structural features of the atoms as such.

The results were in fact in a rather ordinary sort of diffraction pattern, with spots of minimum and maximum brightness (von Laue spots). These patterns appear when an X-ray beam passes through a crystal. The structure of the crystal then alters the path of the X rays so that they create the spotted pattern. Since these patterns are unique for different substances (sort of like an X-ray "fingerprint"), scientists can deduce what type of crystal they are dealing with from its pattern. Thus, these diffraction patterns are one of the main tools for studying crystal structure and crystal defects.

Yet, since he had not yet realized the significance of these results and since the patterns Davisson found did not make any sense according to the classical view of electrons as particles, he thought perhaps the explanation lay somewhere in the atomic structure of the nickel being bombarded. But, unable to make any substantial progress along that line, he abandoned the project.

By 1924, Davisson, tired and frustrated though he was, could not resist making one final effort to make some sense of these results. Working now with Leslie Germer, they hammered away at the problem, stopping only in 1926 while Davisson traveled to England for a vacation and to attend a meeting of the British Association for the Advancement of Science at Oxford. At this point, Davisson was very depressed and was even thinking of leaving physics altogether. He'd invested years of his life and didn't believe he had or would ever accomplish anything. He was starting to believe that all of his research was worse than pointless.

A MIRACLE

What happened then was a scientific Cinderella story. At the British meeting, he was staggered to hear a lecture by the great

Max Born who cited Davisson's work as absolute proof for the quantum mechanical theory! Particles could behave as waves, precisely as Louis de Broglie, one of the founders of quantum physics, had claimed. Davisson and Germer had proved beyond all reasonable doubt one of the most remarkable theories in the history of physics. At the time of Born's talk, de Broglie's ideas on the wave nature of particles were still new and not widely accepted. He had made the remarkable claim too that the wavelength of an electron was a function of its velocity. Indeed, at this point, Davisson had barely even *heard* of de Broglie and his hypothesis about the wave nature of light. After listening to Born, he learned quickly. In fact, he spent the entire return trip to the United States studying quantum theory.

One can only guess at the feelings of a scientist, slowly succumbing to a gnawing sense of worthlessness, who suddenly hears his "pointless" research cited by one of history's greatest physicists as absolutely decisive proof for one of the three most unimaginably powerful scientific revolutions in history (the others, of course, being Einstein's special and general theories of relativity). It can be no surprise that both de Broglie and Davisson were soon awarded the Nobel Prize in physics (1929 and 1937, respectively) for this "useless" work .

By January of that year, he and Germer redoubled their efforts and produced data reconfirming the wave nature of the electron. Their work was confirmed in duplicate experiments the very next year by Japanese physicist Kikuchi in Japan as well as by G. P. Thomson working at the University of Aberdeen in Scotland.

NIELS BOHR APPLAUDS

Soon, Niels Bohr himself, one of the founders of the "old" quantum model of the atom, would also sing the praises of Davisson and Germer, even though it spelled doom for Bohr's "old" 1913 quantum theory. It was destined to be superceded by Schrödinger's 1925 "wave mechanics." A new physics was emerging. In 1927, at the October meeting of the fifth Solvay conference (a series of prestigious conferences held between the wars to bring the world's great physicists together to discuss the latest ideas), Bohr would echo Max Born's words in saying that the findings of Davisson and Germer's work were "very important results which appear to confirm . . . the formulations of wave mechanics."[6] As Germer himself reminisced years later,

> The discovery that electrons behave as waves and produce diffraction patterns when they interact with the atoms in a crystal was made in 1927 by C. J. Davisson and me at the Bell Telephone laboratories, and independently in Britain by Sir George Thomson.[7]

With the death of the Bohr theory, however, physicists would have to get used to nonintuitive ideas. Whatever its difficulties, the Bohr theory was at least intuitively plausible—it could be visualized. With the Schrödinger wave function, one could "picture" subatomic reality only via highly complex mathematical equations.[8]

DAVISSON AND GERMER: LATER WORK AT BELL

Problems of course did remain. According to the principles of X-ray crystallography, one should be able to calculate the wavelength of the electron from the pattern of von Laue spots. However, the data obtained from the experiments did not quite match what was then known about wave mechanics, specifically de Broglie's equations. When Davisson returned to Bell (the name of Western Electric's engineering division had been so changed in 1925), he and Germer continued to study the electron intensities and compared them to their old work. In this way, they hoped to further clarify the relevance of their discovery for quantum mechanics. Finally, on January 7, 1927, they found what they had been searching for after almost a decade—electron behavior that corresponded almost exactly to the predictions of quantum mechanics for the "wave" nature of the electron.

After the work of Pauling, Davisson, and Germer, contributions to quantum mechanics in America very nearly equaled the work of Europe's finest scientists. American physics was coming into its own.

CHAPTER 11

E. O. Lawrence and Nuclear Research

In the early days of nuclear research, even the very greatest scientific minds scoffed at the idea that there might be usable energy inside the atom. The legendary Ernest Rutherford, for one, known for his theory of the disintegration of nuclei into two elements, ridiculed the idea in a *Herald Tribune* article of September 12, 1933, saying, "The energy produced by the breaking down of the atom is a very poor kind of thing. Anyone who expects a source of power from the transformation of these atoms is talking moonshine."[1] Very soon, Rutherford would have to swallow such sentiments. One of the giants of physics who would soon obliterate Rutherford's skepticism was another "Ernest," the eminent Ernest O. Lawrence.

EARLY LIFE

1901 was a great year for science. Ernest Orlando Lawrence was born in Canton, South Dakota, on August 8, 1901, the same year as Linus Pauling and Enrico Fermi. He was the oldest son of Carl and Gunda Lawrence, both Norwegian immigrants. His father, Carl Lawrence, taught history in a Lutheran academy and his mother was a mathematics teacher. Both were devout Lutherans and the couple took care to raise their son in strict accordance with their Christian principles—principles that Lawrence would strictly abide by during his entire life. Lawrence went to school in both Pierre and Canton, but occupied himself after school hours tinkering with wireless telegraphy, tennis, gliders, and other childhood scientific experiments. Later, he would indulge himself in skating and music.

Oddly, he did not perform brilliantly in the lower grades, and first exhibited his vast promise in college. According to scientist and historian Luis Alvarez, "The important ingredients of his success were native ingenuity and basic good judgment in science, great stamina, an enthusiastic and outgoing personality, and sense of integrity that was overwhelming."[2]

Tragedy entered his life early on, when a close cousin succumbed to cancer. Deeply moved by this, Lawrence decided to become a physician, soon entering Saint Olaf College in Minnesota. But he was not happy there and switched after his freshman year to the University of South Dakota. Though he clung for a while to the idea of a medical career, a faculty member, Lewis Akely, continually encouraged him to pursue a career in physics.

GRADUATE WORK

Lawrence received his B.S. with honors in 1922. Immediately he entered graduate work in physics at the University of Minnesota, studying electromagnetism under geophysicist W. F. Swann. It was during this period, though not directly because of Swann, that Lawrence finally gave up his dream of a career in medicine. He received his M.S. in 1923. Swann then transferred to Chicago and, at Swann's suggestion, Lawrence followed him there.

THE MAN BEHIND THE SCIENCE: AN OVERVIEW

Lawrence, for most of his career, showed the most intense concern toward applying science for the benefit of mankind, particularly in the fields of agriculture, medicine, and manufacturing. To a great degree, such humanitarianism was part and parcel of Lawrence's patriotism. He saw the United States as Shangri-La, with a duty to export its values to the rest of the world. During his work on the Manhattan Project, for example, Lawrence more than once grew impatient with those who did not share his vision of a glorious future with the atomic bomb.

Lawrence, though dedicated to his work, could be as sociable as anyone, and loved good conversation with businessmen at the renowned Bohemian Club, which met in a redwood grove outside of San Francisco. He could positively exude politeness and good will. On occasion, he even displayed an awareness of his own limitations. This emerges clearly in a letter of October 3, 1947, to Linus Pauling, where he says:

> Dear Linus, I was delighted to get a copy of your new elementary chemistry text. I expect to read it and profit a great deal as my chemistry certainly does bear brushing up. Of course I am pleased that you mention some of our work and, needless to say, I would not suggest any changes.[3]

And when Pauling won the Nobel Prize in 1954, Lawrence cabled him immediately, phoning Western Union at 3:20 PM on November 5, 1954, saying, "Heartiest congratulations. Hope this belated and so well-deserved recognition will stimulate much additional support for your research work. . . ."[4]

EARLY CAREER

The move to Chicago seemed natural enough considering that such headliners as A. A. Michelson, Niels Bohr, and Arthur Compton were working there. Lawrence did not, however, like the atmosphere despite the presence of such stellar names. The department was hierarchical and tyrannically run, preventing many graduate students from doing physics in their own way. Favoritism was rampant. Disillusioned and wishing to pursue research on the photoelectric effect—the phenomenon discovered by Einstein whereby surface electrons are ejected when light is shone on them—he transferred to Yale in 1924. It was a fortunate move in other ways, for it was at Yale that he met Mary Kimberly ("Molly") Blumer, daughter of a Yale dean. They dated all through his Yale years and married in 1932. They would eventually have six children, Margaret, Mary, Robert, John, Barbara, and Susan.

HEADING FOR BERKELEY

In 1928, he accepted an appointment as associate professor at the University of California at Berkeley. It was at Berkeley that his career would reach its zenith. During his earlier correspondence with the important physicist Raymond Birge, then at Berkeley, Lawrence became convinced that Berkeley would quickly reward fine performances by its faculty, either by advancement in rank, increases in pay, or generous offers of laboratory space.

It was hardly a surprise that he met with considerable hostility when he announced that he was leaving Yale. As he wrote at the time, "The Yale ego is really amusing. The idea is too prevalent that Yale brings honor to a man and that a man cannot bring honor to Yale."

EVIDENCE FOR QUANTUM MECHANICS

In the beginning of his career in California, he continued to work on his early interests, such as the measurement of vanishingly small intervals of time and photoelectricity. He also attacked some of the key ideas in quantum mechanics—still new and controversial in the 1920s. Partly in connection with this work, Lawrence befriended Robert Oppenheimer. In a letter to Lawrence of January 3, 1932, Oppenheimer hints at his [Oppenheimer's] fabled arrogance:

> Dear Ernest, This is an entirely gratuitous little note, . . . I suppose that it is too much to hope that by the beginning of a term the big magnet [for the cyclotron] will be ready; but perhaps by then your contractors will be done. . . . If there are any minor theoretical problems . . . tell them to Carlson or Nedelsky; and if they are stumped let me have a try at them.[5]

An integral part of nuclear research was the cyclotron or "atom smasher," which was, in fact, invented by Lawrence. These were machines that accelerated atomic particles with a burst of electricity, which, in turn, generated a magnetic field that would accelerate the particles to tremendous velocities.

An inspiration for this work was the historic experiment by J. D. Cockcroft and E. T. S. Walton carried out in 1932 at the Cavendish Laboratory at Cambridge University (though Lawrence had had the raw idea of the cyclotron as early as 1929). Their success in splitting an atom of lithium by a proton convinced Lawrence and others that the atom was not, after all, "indivisible"—a prejudice that reached back to pre-Socratic philosophers of ancient Greece. Thus, Lawrence was motivated by nothing more specific than to understand the ultimate structure of matter. (Atomic energy as a source of power was scarcely even on the scientific agenda at this time.) That, in turn, meant probing the interior of the atom with even more energetic, and therefore faster, "bullets" than Cockcroft and Walton had used. Later, in the work on the larger 27- and 37-inch cyclotrons, Lawrence and his group would rely to a great degree on Robert Oppenheimer for theoretical analysis of the results of cyclotron experimentation. (Oppenheimer, arrogant though he was, was such a great physicist that Lawrence still sought his counsel.)

THE ADVENT OF THE CIRCULAR ACCELERATOR OR "CYCLOTRON"

Lawrence, like any good inventor, had happened on the idea for the cyclotron when he was fishing about for new ways to probe the structure of the atom. Soon he discovered a paper by Norwegian engineer Rolf Wideröe. Wideröe, building on some ideas of Swedish physicist Gustaf Ising, had designed a particle accelerator that drove the particle not by a single burst of high-voltage electricity, but with a series of smaller impulses.

Studying this paper, Lawrence realized that the acceleration would be stronger and more efficient if the old "linear" accelerator were retooled as a circular one. Thus, Lawrence designed and built the very first literal or circular cyclotron (the latter term is sometimes used loosely to apply to any "atom smasher"). He selected a circular apparatus simply because it was known at the time that, within a uniform magnetic field, particles could be fantastically accelerated. One critical feature of Lawrence's design was that the motion of the particle was constant and uniform, no matter what its speed (although the diameter of the circle naturally increases at greater speeds).

A strict requirement for a workable cyclotron is that the chamber has to be as near to a perfect vacuum as possible, lest the particle lose speed by colliding with gas molecules. The cyclotron itself is therefore merely a large hollow tube, divided into two semicircles called "dees." (To get more speed, larger dees are needed and of course more power.)

When a charged particle is placed inside the tube, it is attracted to the dee having the opposite charge of the particle. Thus, a proton, which is positively charged, is attracted to the negative dee, and is therefore accelerated. It then travels through the circular tube, finally being ejected out of the tube opposite the point of entry. Technicians then reverse the voltage so that the proton steers for the opposite dee—now bearing a negative, rather than a positive charge. Thus, it reenters the first dee at an even higher speed and, therefore, in a wider circle. The point is that once the particle has been ejected the first time as discussed above, it has already been accelerated and therefore energized considerably. Thus, the particle, already energized from its first "trip" around the cyclotron, upon reentering the device is given a second burst of energy. This accelerates and energizes it still more. (This is sort of like giving a child a ride on a small, hand-driven merry-go-round found in any amusement park; you give it a push and the merry-go-round accelerates. Give it another push, and it goes even faster.) With each cycle, the proton is forced to even greater velocities and ever-widening circles until it is at the perimeter of the cyclotron. Once this perimeter is reached, it is at its maximum speed and the staff then "aims" it at its target.

By the end of 1936, there were at least 20 cyclotrons elsewhere in the world, and they were improving. It was at about this point in history that cyclotron research really became a global phenomenon. Even Lawrence's smaller 37-inch cyclotron—still in use in late 1937—was looking less amateurish than the earliest versions of the cyclotron. (The term *cyclotron* is an extremely general term applied to any "atom-smashing" apparatus that accelerates protons in a circular fashion. It is not the "brand name" of any given "model.") It had a gasket groove or ring running around the outside circumference to make it airtight, as well as surfaces that were clearly machine-tooled. The deuteron energy had also now soared to 8 MEV, i.e., 8 million electron volts, a measure of the speed, and therefore, the energy of the particle being accelerated.

Of course this took money—big money. In the spring of 1940, the Rockefeller Foundation gave Lawrence over $1 million for the construction and maintenance of his 100 million-volt cyclotron. Later he would be assisted by the Macy's Foundation, the WPA (Wage and Price Administration), and other organizations. Although Lawrence had argued for the great value of this work for medicine in treating cancer, the great impetus for such largesse on the part of the government was, of course, the war and the need for beating Germany in the race for an atomic bomb.

AN ILL-BEGOTTEN THEORY

Early in his career, Lawrence latched onto the idea that deuterons, or the nuclei of a heavier version of the element hydrogen, or "deuterium," were fragile things—easily disrupted under ordinary lab conditions. On December 28, 1933, he wrote Soviet-American physicist George Gamow, well known for his work on radioactivity, referring to "The experimental evidence that the deuton disintegrates. . . ."[6] Soon, he theorized too that the nucleus of deuterium or a "deuteron" (earlier called a "deuton" by Berkeley chemist G. N. Lewis) could be easily busted into its component nucleons (any particle located inside the atomic nucleus) through repeated collisions. The practical significance was immeasurable, as it was still another indication that atomic fission or atom splitting was possible and so, therefore, was the atomic bomb.

And even when the experimental work of men like John Cockcroft, the first scientist to actually split an atom of lithium into two atoms of helium in 1932, did not bear out this *particular* alleged example of "atom smashing" Lawrence persisted, merely because he had immense faith in his own scientific judgment and his own experimental results. Thus, 1934 was a difficult and troubling year for him. Replying to Cockcroft's wholesale rejection of his theory, Lawrence wrote him a letter on January 12, 1934, which, by the lofty standards of scientific politeness, can only be considered "cranky." "It seems to me that you are hardy justified in feeling that the evidence obtained by you so far is against the interpretation of the break-up of the deuton. . . ."[7]

Also in July of that year, Lawrence's trusted colleague Livingston left the laboratory for Cornell (though the reasons he did so are unclear). However, Edwin McMillan stayed on to work further with Lawrence, eventually winning the Nobel Prize with American chemist Glenn Seaborg for their discovery of the element plutonium, and taking over Lawrence's post as the head of the laboratory. Eventually, of course, Lawrence did have to abandon the idea that he had smashed a deuton, when the evidence against it by Cockcroft, Oliphant, Rutherford, and others proved overwhelming.

EARLY WORK IN NUCLEAR PHYSICS

Soon, Lawrence turned his attention to the field of nuclear physics. Though the field had existed for some years, it was relatively unprestigious until Rutherford split the nucleus by bombarding it with helium nuclei emitted by radium, jarring even his own infamous skepticism about the amount of energy contained in the atom's nucleus.

Still another critical historical event was the work in 1938 of Hahn and Strassmann on nuclear fission. This was taken further by Otto Frisch and Lise Meitner[8] who proved that the uranium nucleus could be split by an energetic hit with a neutron. On February 9, 1939, Lawrence wrote his colleague Cockcroft as follows: "We are having right now a considerable flurry of excitement following Hahn's announcement of the splitting of uranium."

Lawrence immediately realized that this discovery could lead to the production of an atomic bomb and that Germany would probably be working on one as well. That thought terrified him. Lawrence had mistrusted the Nazi regime as early as 1929 when Hitler appointed Himmler as "Reichsfuhrer S. S.," following this by the naming of a few other close friends to important posts. Even before Nazi atrocities were fully documented, Lawrence had another cause to dislike the German oligarchy. Rutherford, in private correspondence, had told him of the difficulties scientists were facing in Germany. He wrote Lawrence in the fall of 1936:

> . . . This state of affairs in Nazi-land is rather amusing, and when some of our men from the Cavendish wished to visit Berlin to see Debye's laboratory, he wrote to Cockcroft that official permission

would have to be granted by the Government before he could admit
them!

Lawrence replied to Rutherford on February 11, 1937:

> Your account of the state of affairs in Germany is almost unbeliev-
> able. One would think with such a scientific tradition the German
> people could not adopt such an absurd course of action in scientific
> affairs.

By 1943, interest in nuclear research would reach manic proportions.

LAWRENCE MEETS AN AMATEUR

Lawrence's supply of money did not match his natural fund of
talent, and he needed help. Lawrence found this aid from somewhat unusual sources. In 1939, he first met Alfred Loomis, whom
physicist and historian Alvarez has convincingly argued was the
last of the important 19th-century-style scientific amateurs.
Loomis actually did credible science almost literally in his "backyard," despite not having a university post and without outside
sources of funding. Though trained as an attorney, Loomis was
almost as knowledgeable as any scientist in physics. He fully
appreciated the costs and politics of high-energy science and
empathized with Lawrence's financial plight. Soon, he managed
to persuade the Rockefeller Foundation to give Lawrence the $2.5
million he would need. By 1939, Loomis himself joined Lawrence's
team, both as a financial mentor and as a scientific advisor, to
apply science to wartime needs. Soon he was working along with
Lawrence on a microwave radar aircraft detection system.

At the end of WWII, Lawrence paid tribute to Loomis:

> He had the vision and courage to lead his committee, as no other
> man could have led it. He used his wealth very effectively in the way
> of . . . making things easy to accomplish. . . . I am perfectly sure

that if Alfred Loomis had not existed, radar development would have been retarded greatly, at an enormous cost in American lives.[9]

CYCLOTRON RESEARCH ESCALATES

In 1933, the distinguished American engineer Malcolm Henderson and Telesio Lucci, a former Italian Navy officer, joined Lawrence's research staff. Their expertise in engineering and electronics proved invaluable for contriving data-recording instruments and magnetic circuits, as well as helping repair the cyclotron, which broke down quite frequently. With their aid and larger magnets, Lawrence and his team were soon able to amplify their cyclotrons so that that their power reached into the billions of electron volts.

After he'd built his own cyclotron, Lawrence did not hesitate to help others interested in building them. Indeed, the very first request Lawrence ever received about building a cyclotron on foreign soil came from the renowned Frédéric Joliot who, with his wife Irène Joliot-Curie, was soon to win the 1935 Nobel Prize in chemistry for synthesizing new radioactive elements. Joliot wrote Lawrence on June 14, 1932, saying he had been following Lawrence's work: "Votre travail me parait remarquable, faire avec de tels rayons sont d'un grand interet. [Your research intrigues me, as do the lines of inquiry that such rays make possible.]"[10] Lawrence replied on August 20, saying to Joliot that he might be able to assist him in obtaining some of the necessary equipment. Later he would aid a number of scientists in building similar machines as well, including Chadwick, who had recently acquired immortality by capturing the 1939 Nobel Prize for the discovery of the neutron, which he announced at the Kapitz Club of the Cavendish Laboratory in England.

The work in nuclear research was not without controversy; popular newspaper accounts, stressing the awesome voltages involved, made them seem like Hollywood "B movie" experiments. But scientists were only mildly perturbed. Human nature being what it is, the various cyclotron laboratories were soon racing to outdo one another in producing a bigger and better atom-smashing apparatus.

ATOM SMASHING AND NUCLEAR CHEMISTRY

As the design of the cyclotron improved, Lawrence could better control both the direction and the energy level of the accelerated particles. With these improvements, the field of nuclear chemistry was born. Physicists were now able to bombard the nuclei of virtually any substance in a controlled manner. Usually, the assault on the elements would break them into isotopes of still other, well-known elements.

Once in a while, however, the collisions produced elements never seen before in nature. Once the work was begun, researchers found many elements that Lawrence and others had postulated, but that had never been seen. For example, Lawrence and his co-workers were able to create and identify radioactive iodine, sodium, and many other elements.

All of these "artificial" radioactive substances lost their radioactivity in a matter of days and it was now clear why light radioactive elements had never before been found in nature. While they must have existed at some time long past, their life spans were so short that they had all disappeared long ago. The only radioactivity still left on Earth was located in the very long-lived elements like uranium, thorium, and radium.

LAWRENCE'S LARGESSE LEADS TO STILL NEW UNDISCOVERED ELEMENTS

Lawrence was not the only one interested in undiscovered elements. The celebrated Italian physicist Emilio Segrè, working with artificially created radioactive material sent to him by Lawrence, began studying a new and different nuclear "building block"—the meson—so called because its mass was less than a proton, but far greater than an electron. Soon, theorists discovered the existence of another particle postulated along ago—the antiproton—a particle that in all respects was identical to the proton except that it had a negative rather than a positive charge.[11]

Segrè, working with his assistants, discovered many of the theorized and predicted, though still missing elements. One particularly peculiar one appeared when he bombarded molybdenum. After the radioactive residue was separated out, it behaved similarly to the known element rhenium. But it was not rhenium. With further work, they proved that it was yet another of the missing elements—number 43. Number 43 was an entirely synthetic element, and therefore not found naturally in the Earth's crust. Segrè named the missing element "technetium," from the Greek word meaning "artificial." (Segrè was also part of a team that found the 92nd element, "astatine," in 1940.)

This was possible only because Lawrence's work told them how to alter the numbers of subatomic particles in the nucleus. It is these very particles that make a substance the kind of element it is. If you add to or subtract from the particles in the nucleus, you have changed it into another type of element.

THE UNFATHOMABLE POWER WITHIN THE ATOMIC NUCLEUS

The decades of work by Fermi, Lawrence, Segrè, and many others working in nuclear physics gradually yielded sufficient data to tell physicists how much energy was required to hold the nucleus together. By the same token, they reasoned, this was the amount of energy that would be released if the nucleus were to be taken apart. At around this time, the truly awesome possibilities of atomic power were beginning to be appreciated by the scientific community.

But the physicists working in this area, such as Segrè, Edward Teller (who later helped develop the hydrogen bomb), Fermi, and many others, discovered something fantastic: The simplest nucleus was hydrogen—only one proton. The next simplest was "heavy" hydrogen—one proton and one neutron. The sum of the masses of these two particles as free particles was 2.0171 units. The mass of the two in combination as the "heavy" hydrogen nucleus was 2.0147 units, which meant that the combination was lighter than the sum of the separate particles by 0.0024 units. Another way of saying it is that 0.0024 units of mass had been *annihilated* when the two came together.

This number—the slight discrepancy between the two weights—was one of the most revealing numbers in nature; it was the *direct measure of the energy required to bring a neutron and proton into combination*. In this way, the world had experimental proof of Einstein's equation of mass with energy. This kind of nuclear bookkeeping was then applied to helium, which consists of two protons and two neutrons. The process was continued for every element known, and in every case the nucleus was lighter than the sum of the particles it contained. It was found that the nuclei hoarding the greatest amounts of energy—and which were, therefore, the most stable—were those of intermediate weight. The heaviest elements and the lightest proved to be the most unstable in theory, as they were in actual fact.

INSTABILITY IN NATURE

Things grew more complex when, besides the unknown nuclear force that held the particles together, scientists realized that there was yet *another* force tending to disrupt them. This was the simple law of electrical repulsion. The larger the nucleus, the more particles it contained and the more important became this disruptive force. The current theory further stated that there would be a critical size to the nucleus beyond which this force of disruption would be even more powerful than the force holding the particles together. Therefore, any nucleus with a diameter larger than this would tend to come apart on its own.

More calculations showed that this limit was reached by the most massive elements—uranium and thorium. Because they were on the verge of disruption, they were highly unstable. Every so often, said the theory, one of these nuclei would have to eject enough mass from itself to come down to a safer size. These ejected particles were called the "radioactive emanations." By 1939, "natural" radioactivity—of radium, uranium, and thorium—was at last seen to be nothing more than this cataclysmic self-regulation of unstable nuclei.

In 1939, over 45 years of study of radioactivity had been completed; and at last a true semblance of order seemed about to emerge. There was the sense that all of the Great Truths of physics were now known—that same sense of serenity that Michelson had felt earlier in the 19th century—before Becquerel, Roentgen, and J. J. Thomson had pulled the foundations out from beneath what was then canonical. But this time there was one big difference: in 1939, while physicists were operating under correct information, they just did not fully appreciate the apocalyptic implications of their information.

Soon, this work on nuclear fission led to the 1939 Nobel Prize for Lawrence. Naturally, the receipt of the prize greatly accelerated the cyclotron work. Lawrence immediately started hunting down more money to build an even grander and more powerful 100 million-volt cyclotron—a difficult task. Lawrence's quest for larger cyclotrons was understandable; though he had no specific goal in mind, he knew well that the more powerful a machine he had, the more he could probe the secrets of the nucleus (in much the same way that astronomers want larger and more powerful telescopes). In the 1930s, there was no Atomic Energy Commission money and no Manhattan District to aid needy projects. But he persisted and, somehow, managed to get by with private donations. Not quite anticlimactically, by April 1940, both the Regents of the University of California and the Rockefeller Foundation caved in under Lawrence's pressure and donated $1.5 million total for more cyclotron research which Lawrence used to construct still more powerful "atom smashers."

By 1940, Lawrence's theoretical work was leveling off, and he started devoting more time to administrative and public interest work. In the early stages of World War II, the United States and its allies needed antisubmarine technology. Nazi forces were wiping out shipments of food and spare parts that the United States had been sending to England. To solve that problem, Lawrence and his group established an underwater sonar laboratory in San Diego.

WORK ON THE MANHATTAN PROJECT

And at the same time, while he was at the University of California at Berkeley, Lawrence began working on still another critical military problem. The Manhattan Project, the secret government program that developed the first atomic bomb under the scientific leadership of J. Robert Oppenheimer, had run into a serious difficulty: how to separate U-235 from U-238. In order to create an atomic bomb, it was necessary to be able to separate these two isotopes of uranium, inasmuch as only the atoms of the lighter isotope were practically "fissionable," or easily split into smaller atoms. Thus, Lawrence began converting the cyclotron into a mass spectrometer to solve this problem.

The principle of the spectrometer was in some ways similar to the cyclotron, in that it too used electromagnetic fields. The spectrometer, however, would send particles selectively in different directions, depending on their mass. Because isotopes differ in mass, the U-235 isotope nuclei would follow a different trajectory than the heavier uranium isotope, and could therefore be collected separately.

Although the spectrometer method was eventually replaced with a faster, less expensive, and, therefore, more efficient gaseous diffusion system, the spectrometers soon built at Oak Ridge, Tennessee, by researchers on the newly established Manhattan Project proved invaluable in producing usable uranium. However,

Lawrence and his staff at Berkeley still continued to produce much of the uranium needed for the Manhattan Project with the methods described above.

THE AFTERMATH OF WWII

After the war, the scientific community in the United States slowly returned to "pure" research—unencumbered by the necessity of gearing their work for the war effort. Lawrence, however, continued to work on nuclear military capabilities, lobbying for another research installation at Livermore, California. That facility, in turn, made possible the subsequent development of the even more deadly hydrogen bomb.

MEDICAL APPLICATIONS

Yet the advent of the atomic age was not all designed and utilized for weapons of mass destruction. It is to Lawrence's credit that he envisioned far more peaceful and life-saving functions rather than life-taking uses of atomic power. With the advent of the cyclotron era, it became possible to solve a problem that Lawrence had encountered earlier. In the past, in order for him to work with radioactivity, he'd been forced to borrow such substances from hospitals for only limited periods. Now they could be created at will in the laboratory. Working with his younger brother, John Lawrence, who was a physician as well as chief of the Berkeley Biophysics Laboratory, the pair managed to treat a variety of malignant sarcomas (a type of cancer often found in connective tissue) with the newly created isotopes (varieties of radioactive substances with the same number of protons in the nucleus, but of slightly different mass). They did this by bombarding the cancer with beams of neutrons. Of course it was dangerous. In order to bombard tumors far inside the body, they had to use beams so intense that they thought they might risk burning healthy tissue, since that is what happened when X rays were used in cancer treatment. But bombardment with neutrons destroyed cancerous tissue far more efficiently than X rays without such risks. Although the side effects, including nausea and general debilitation, were quite severe, the treatment was very promising. In a letter to Ernest Rutherford in March 1936, Lawrence describes their work:

> . . . I believe in my last letter I mentioned that we have been carrying on experiments on the biological action of neutron rays. . . . My Brother, Dr. John H. Lawrence of the medical faculty of Yale University, has been out here studying the effects of neutrons on a certain malignant tumor called mouse sarcoma. . . . He has compared the lethal effects of neutrons and x rays on the tumor and on healthy mice and has very impressive evidence that this malignant tumor is relatively much more sensitive to neutron radiation. . . . I am sure that it will not be long before neutrons will be used in the treatment of human cancer. . . .[12]

In September 1934, Lawrence managed to create a radioactive isotope of sodium by bombarding ordinary sodium with deuterons. Not only did it turn out to have fewer poisonous side effects than radium, but with a half-life of only a few hours, it lasted long enough to wipe out cancer cells, but not long enough to do any irreparable harm to healthy ones.

Commenting on this in a letter to Rutherford, Lawrence says: ". . . In this country medical research receives generous support, and it was the possible medical application of the artificial radioactive substances and neutron radiation that made it possible for me to obtain adequate financial support."

NEUTRONS BETTER THAN X RAYS IN TREATING CANCER

After his work with his brother in treating mouse sarcoma, Lawrence turned to a very personal and daring application of radiation therapy. In 1937, the two brothers used radiation to treat their own mother. Though already considered "terminally ill" with cancer, the new radiation therapy was successful enough to let her live another 20 years. With expected improvements in technology, and a better understanding of dosages, radiation therapy is still in use today.

Lawrence won a very wide variety of honors in his long and distinguished career, including the Nobel Prize in 1939, "for the invention and development of the cyclotron and for results obtained with it, especially with regard to artificial radioactive elements."

In addition to the Nobel Prize, he was also awarded the prestigious Hughes Medal, given by the Royal Academy of Sciences. Though he gradually slowed his pace, he never stopped working. In addition to numerous military contributions. Lawrence also designed an improved television tube called the Lawrence Chromatron, now manufactured by Sony.

Lawrence died of colitis in Palo Alto, California, on August 27, 1958, but his work is one of the finest examples in history of the power of science to affect the human condition. It is no exaggeration to say that his work revolutionized the treatment of cancer and, because of his invention of the cyclotron, he greatly accelerated our understanding of the inner workings of the atomic nucleus.

CHAPTER 12

Carl Anderson and Antimatter

Since ancient Greek times, philosophers and scientists have sought the ultimate building blocks of nature. For a while and throughout the 1950s, many thought they'd found them all. But, just as drama and excitement are among the great qualities of physics, so is frustration. And since the early 1960s, every time physicists began to think they'd found all of the basic particles, a new one would appear to tease them. As former presidential science advisor Robert Bacher told me in his office at Caltech, "it seemed like a bottomless pit."[1]

The era of modern particle physics was beginning. But, although the rapid and massive proliferation of particles is relatively recent, the story really begins much earlier.

In 1927, the new science of quantum mechanics was celebrating only its second birthday. Although the general ideas were fairly widely accepted, they still bothered some theorists. The principal reason was that it denied some long-cherished ideas of how nature worked. The Heisenberg uncertainty principle, for example, in claiming that there was no certainty in scientific predictions, violated the old idea of Newton and French philosopher and mathematician René Descartes which stated that it was possible for science to achieve absolute knowledge about the workings of the cosmos. (Even Einstein was in the grip of this old notion and fought strongly against the new uncertainty principle.) However, in the late 1920s, Cambridge theoretical physicist P. A. M. Dirac began working on an extension of quantum mechanics that he hoped would be consistent with Einstein's theory of special relativity. In February 1927, he published a ground-breaking paper. In essence, it was a version of the famous Schrö-dinger equation of quantum mechanics, but which took Einstein's theory of relativity into account. (The Schrödinger equation describes how subatomic particles will behave in any force field.) A year later, Princeton physicists Pascual Jordan and Eugene Wigner applied these principles and concluded that only one electron could occupy a single quantum state—precisely in line with Dirac's ideas about electrons and positrons. Dirac's theory, as noted above, was supposed to predict how particles would behave in a force field, and this was one of the predictions: his theory predicted that a negatively charged electron, being attracted by the positively charged nucleus of an atom, would have to be the sole occupant of a given quantum state. That, as it turns out, was true. (The idea that no two electrons could occupy a single quantum state is, intuitively, sort of like the idea that no two persons can occupy the same place at the same time—phone booths being a good example.)

However, Dirac was dismayed to find that his new equations, which were intended to describe the dynamics of the electron, were also causing considerable trouble. The equations he used for describing the energy of the electron in his new theory turned out to have both positive and negative solutions. In other words, his theory was saying that they could have both negative and positive "energies."

But what was a "negative energy"? As he put it in his Nobel Prize address:

> There is another feature of these equations which I should now like to discuss, a feature which led to the prediction of the positron. If one looks at Eq (1), one sees that it allows the kinetic energy W to be either a positive quantity . . . or a negative quantity . . . Now in practice the kinetic energy of a particle is always positive. . . .[2]

And further:

> Thus in allowing negative-energy states, the theory gives something which appears not to correspond to anything known experimentally. . . .[3]

A PHILOSOPHICAL PROBLEM

While this prediction was consistent with Dirac's own theory as described above and, in fact, did not contradict any established principles of physics, most physicists resisted the idea with great force, simply because it ran counter to our ordinary intuitions about reality. Even worse: since Einstein had shown that energy and matter were equivalent, there would have to be such a thing as "negative *mass*." Dirac, of course, appreciated this problem and at first suggested that such "negative energies" were mere artifacts of the mathematics, rather than literal descriptions of the physical universe.

OCEANS OF ENERGY

Then he happened upon a brilliant new idea: since the "negative" energy states appeared inevitable, he had to find some way for his theory to incorporate them. So, he now supposed that the universe really was filled with what he called "negative energy states," sort of like a "hole" in a region of space, consisting of this mysterious "negative energy." (Imagine that throughout the Pacific Ocean there were spots where if you stuck your hand in the water, it would come out *dry*.) And, he decided that each negative state was filled with a single electron. As he says, "We then have a sea of negative energy electrons, one electron in each of these states. . . ." And, since their existence "fills" the negative energy states by virtue of the Pauli exclusion principle (no two electrons can be in one state of motion, the universe is stable. With a negative-energy electron already in this "hole," nothing else, such as a positive electron, can fall into it.

Here I think a brief digression is necessary. Something should be said here about the idea of "detection" when speaking of such almost infinitely small entities as electrons. For starters, there is no microscope, even the electron microscope, that can come anywhere near to "seeing" an electron. In fact, it would be a mistake to think of most subatomic particles as "things" like breadboxes or erasers, or even as molecules or bacteria. At this level of size, our ordinary concepts of "things" as extended in space do not necessarily apply. According to some speculations, for example, an electron has no "size" at all! It is infinitely small. If that seems "illogical," the answer is that our ordinary ideas of "logic" do not apply at the subatomic level. (This is reminiscent of the Zen Buddhist's response to the Westerner who finds Buddhism "illogical." The typical Zen answer is that Western "logic" simply does not apply to Zen.)

Dirac then pointed out that inevitably light would occasionally slam into the electrons in the "negative energy sea." Like a cosmic Geritol tablet, this collision would increase their energy so much that they could "hop out" of the sea. They would then become "positively" energized electrons.

Of course, by jumping out of the sea, these souped-up electrons would leave a "hole" in the sea of "negative energy" (just as pulling a tree out of the ground leaves a hole in the Earth). Dirac decided that this "hole" could in fact be just as easily thought of as a particle—an electron, but with a charge *opposite* to that of a standard electron. And since an ordinary electron has a negative charge, this "hole/particle" would have a *positive* charge. In short, Dirac's theory predicted the existence of a positively charged electron.

Of course, all of this was theoretical speculation on Dirac's part, albeit speculation well-founded in scientific theory. But there was, as yet, no actual proof of the existence of such positive electrons.

Dirac was hardly overjoyed by his theories, however. The entire "ocean" idea was so speculative that it bordered on medieval theology. At first, he tried to follow Ockham's razor, a principle in science that forbids making explanations any more complicated than necessary. So initially he wondered whether these particles could simply be protons, which were already well known.

However, mathematician Hermann Weyl soon showed that ordinary protons wouldn't fit—they were too massive—in Dirac's "holes," so Dirac had to accept the idea that his holes would have to contain particles having a mass no greater than that of an electron. He drew some comfort from the fact that although his postulation of such electrons was difficult to confirm for the reasons given above, at least they did not *conflict* with observation. But of course, the same is true of the supposition that invisible green men and Italian-American angels exist. So it is not surprising that many physicists had philosophical differences with Dirac's view.

ANOTHER DILEMMA

Many were also troubled over another apparent glitch in Dirac's theory. Powerful though it was, physicists realized by 1927 that Dirac's treatment of the electron was incomplete. Quantum mechanics was already proving to be a powerful tool for describing how subatomic particles as well as radiation interact with one another—the forces involved, the different possible energy states of the atoms involved, the radiation given off, etc. Dirac had given a quantum mechanical description of the electron all right, but he had done this only for the radiation and not the "particle" part of the electromagnetic field.[4] In other words, the favored theory at the time as well as today was that an electron could be regarded as both a "wave" and a "particle" (another instance, by the way, of our ordinary everyday conceptions of the world breaking down at the subatomic level). Yet Dirac had essentially ignored altogether the "particle" aspect of the electron. Why he did so is not clear.[5] Probably, however, he saw that this led to the bizarre and almost unbelievable concept of *antimatter*, in this case an antielectron. As some commentators have suggested, his fabled courage may simply have deserted him at this point. That is why he refused to consider the idea of an antielectron.

But this disturbed many, as it appeared contrary to the spirit of the gradual unification of all physical phenomena, which had been ongoing since Maxwell unified electricity and magnetism back in the 19th century. Maxwell had shown that electricity and magnetism, which had previously been thought of as two *different* things, were really merely two aspects of one and the same thing. (Is a pot of ice in winter a "different" thing from the pot of water in summer? No, it's really the same substance, namely water, in both cases. It merely appears to us in two different forms. In this same way, electricity and magnetism are really the same thing, appearing to us in different ways.)

And this effort fits into a more general pattern. Physicists and philosophers ever since ancient times have tried to show that what appear to be different things are really just different manifestations of the same thing. So, in 1929, Pauli began efforts:

> . . . to connect, in a contradiction-free manner, mechanical and electrodynamic quantities, [i.e., particles] on the one hand, and radiation. . . . on the other, and to treat them from a unified view point. . . .[6]

(Further success in unification had to await the work of Richard Feynman, Tomonaga, and Schwinger in the 1960s, whose justly renowned work removed many of the inconsistencies in quantum electrodynamics that neither Jordan, Dirac, Pauli, nor Heisenberg were able to solve.)

On the brighter side, even though the negative electrons weren't detectable, Dirac's theory did predict other phenomena that *would* be detectable. For instance, the *positively* charged electron described above might be locatable. But all this talk was still highly theoretical and speculative. It was one thing to predict the existence of positive electrons—known as positrons—but quite another to actually find them in nature. However, the prospects for accomplishing this brightened when several other scientists offered some powerful mathematical reasons for believing that all that would be required for a positron to be created is an energy at least double that of the mass of the electron.

And find them they did. Soon, all of these theoretical worries faded when Carl Anderson began his studies of cosmic rays and, in so doing, made his fabled discovery of the positron—confirming exactly what Dirac had predicted.[7]

HELP FROM THE HEAVENS

The solution to Dirac's *angst* came ironically from an apparently unrelated source: cosmic rays. The study of such rays is itself a fascinating piece of history; the first person to actually observe the track of a cosmic ray, though he didn't realize what he was seeing at the time, was Soviet physicist Dmitry Skobeltzyn in 1923. Working in his laboratory in Leningrad, he knew that gamma rays, which come from cosmic rays, could dislodge electrons from the outer or "valence" shells of most atoms. Logically, then, a theorist might hope to see the tracks of such "evicted" electrons in a cloud chamber of sufficient power and sophistication (since such chambers detect cosmic rays). A cloud chamber, familiar in many high school laboratories, is merely a metal box filled with gas and moist vapor. Particles traveling through the vapor will leave tracks in it; often, a trained observer can deduce from the path and size of the track what particle must have caused that track.

Skobeltzyn wanted to study only the electrons being produced *within* the cloud chamber. So, in order to screen out stray electrons evicted from other sources, such as the cloud chamber itself, Skobeltzyn surrounded the inner detecting area of the chamber with a magnetic field. In that way the field would draw off possible spurious electrons from outside sources and keep them from interfering with his measurements. With this apparatus in place, he began taking a series of photographs of moisture trails in the chamber in the years 1927 and 1928.

Oddly, some of the tracks were very nearly straight. He believed that this could mean only one thing: whatever particles were leaving these "tracks," they had to be much more massive than an ordinary electron. If they had the mass of ordinary electrons, the curves would have been bent much more than they were, since electrons are so light. (The heavier a particle, the more difficult it is to alter its path, just as it is more difficult to shift the path of an onrushing truck in comparison with an onrushing bicycle.)

However, Skobeltzyn realized that there could be another explanation of the nearly straight tracks. If he assumed that the particles following these weird paths were actually electrons of ordinary mass, they would also be very hard to deflect if they were traveling at unusually *fast* speeds. So, he next theorized that the tracks could be due to electrons that had been knocked out of their atoms, and which were also traveling with prodigious velocities—a reasonable assumption considering the immeasurable power in a cosmic ray.

In fact, using these very ideas, American physicist Carl D. Anderson managed to unveil the evidence for Dirac's shrouded "positive electron."

ANDERSON: EARLY LIFE

Anderson was born in New York City on September 3, 1905, the year of Einstein's special theory of relativity and the completion of the legendary observatory on Mt. Wilson. His father was a modestly successful restaurant entrepreneur in Los Angeles and both of his parents were Swedish immigrants. As his mother said of her son, "If he has special ability, I don't know where he got it. . . ." (Indeed, his talent was not evident at first: he struggled mightily in Robert Oppenheimer's course in quantum theory at Caltech.)

While the family was never wealthy, they saw to it that their son was educated under the best possible circumstances, and Carl was soon attracted to the fledgling California Institute of Technology. He graduated in 1927, with a bachelor of science degree in engineering and physics. After this, he limited himself entirely to physics and received his Ph.D. in 1930, also from Caltech. His genius eventually became evident to his colleagues. In an early letter of August 31, 1930, from Millikan to Oppenheimer, Millikan, in trying to entice Oppenheimer to join the Caltech faculty, uses Anderson's work as a lure:

> . . . Carl Anderson is alive to all possibilities . . . through his present cloud chamber, the largest in existence, [and also] through flights to the ceilings attained by B-29's . . . carrying at that altitude cloud chamber instruments.[8]

In 1933, Caltech appointed him as an assistant professor of physics and in 1939 elevated him to full professor.

His earliest work was in the field of X rays and he completed his Ph.D. dissertation under Millikan on the emission of electrons from gases by X-ray emission. As with many scientists, his early years were inimitable, culminating in the discovery of the positron in 1932. For this, he received the 1936 Nobel Prize. This discovery represented the pinnacle of work begun earlier with Millikan and the cloud chamber on the nature of cosmic rays.

ANDERSON AND THE MILLIKAN–COMPTON CONTROVERSY

In 1930, Anderson was a graduate student under Millikan. Millikan asked him to remain at Caltech to help with the comic ray studies. Anderson agreed, although he had some concerns over Millikan's religious view of science. Anderson also had recently designed and built the most powerful cloud chamber to date, which was many orders of magnitude greater than the first prototype.

TENSION WITH MILLIKAN

At the same time, Millikan was embroiled in his infamous cosmic ray controversy with Arthur Compton. And despite overwhelming evidence to the contrary, Millikan still regarded cosmic rays as photons, rather than charged particles. (Photons are essentially the "particles" making up ordinary light, and are uncharged.) Naturally, Millikan was shocked to find that the cosmic rays entering the cloud chamber contained approximately equal amounts of positively and negatively charged particles. It was at this point that Anderson first speculated that the positive particles found in the chamber might be protons—also evicted from atoms by the intense power of cosmic rays.

This is where Anderson's genius triumphed over Millikan's stubborn defense of his moribund "photon" theory of the nature of cosmic rays. Although Anderson could hardly dare challenge the paranoid Millikan too directly, the few challenges he did issue resulted in heated disputes. When he suggested, for example, that the tracks might be due to electrons, Millikan passionately rejected the idea.

Desperately anxious to prove he was right, Anderson placed a lead plate on the chamber to slow down the particles as much as possible, making the deviations in the chamber that much more obvious and measurable. Anderson continued by tracking cosmic rays in the cloud chamber which curved in a path whose shape was identical to the tracks of electrons, but pointing in exactly the opposite direction.

His observations were nothing short of epochal; the mysterious tracks had definitely been caused by an electron. At the same time, the path of the track proved almost beyond question that the particle making the track must have a mass about the same as an electron—but was carrying a positive, rather than the usual negative charge.

Part of Anderson's success in measuring cosmic rays resulted from a judicious use of some colossal magnets, to improve the efficiency of routine measurement procedures for the rays. Again at Millikan's urging he began using the cloud chamber to study the production of photoelectrons by X rays.[9] After he found the tracks of positively charged particles, he first tried to prove that they were nothing but ordinary protons, particles much more massive than electrons, but without success. It was at this point that he began to suspect that they might be the positive electrons predicted earlier by Dirac's theory.

In September 1932, despite more misanthropic emanations from Millikan, he published his finding of the positive electron. Linus Pauling remembered this period well and reminisced about it during an interview in 1964:

> I knew Carl Anderson well: he was a Junior Travel Prize winner. Then I knew him as a graduate student when he was here [Caltech]; this study of cosmic rays with the Wilson Cloud Chamber . . . was something I suppose Millikan had suggested to him. . . . The only thing that I remember about the discovery of the positron that might have some special interest is that E. T. Bell, who is now dead, told me just about the time that Carl Anderson published this paper that Carl mentioned to him that he had a pair of tracks and it looked as though one of them was a positive electron but that he thought that perhaps Professor Millikan didn't want it published or something like that. . . .[10]

OTHER SOURCES OF POSITRONS

Although such particles are extremely rare, they are now seen not only in cosmic rays, but in such energetic disruptions as exploding stars and the immensely powerful collisions between subatomic particles that occur in giant particle accelerators. Of course, Anderson had other detractors and skeptics besides Millikan. The fabled Niels Bohr thought his view preposterous. Nor did he fare better in England. Anderson's 1932 lecture at the famed Cavendish Laboratory was met by jeers from the audience, many being convinced that the "positron" was the result of a simple confusion.[11]

DIRAC TRIUMPHANT

Everywhere he spoke, Anderson had to remind people that precisely such a particle had been predicted by Dirac. Eventually, most scientists were finally convinced that this was indeed the antielectron predicted by Dirac's theory. Anderson soon dubbed it the "positron," and the Nobel Committee awarded still another prize to an American experimental physicist.

The vast new field of "particle physics" had truly begun. In a letter to his brother Frank, Robert Oppenheimer speaks hopefully about future prospects of physics after Anderson's, as well as other discoveries:

> Lawrence's things are going very well; he has been disintegrating all manner of nuclei. . . . We have been running a nuclear seminar, in addition to the usual ones, trying to make some order out of the great chaos. . . . We are supplementing the paper I wrote last summer with a study of radiation in electron–electron impacts, and worrying about the neutron and Anderson's positively charged electrons. . . . I take it that there will be a lull in the theory for a time; and that when the theory advances, it will be very wild and very wonderful indeed. . . .[12]

Not only was this discovery world-shaking because it was the first time a positive electron had been found in nature, it was also the very first time anyone knew that there were "nuclear" particles that did not come from the nucleus! (At one time, electrons were believed to exist in the nucleus of atoms.) But where *did* they come from?

FURTHER VINDICATION OF ANDERSON AND DIRAC

Subsequently, physicists Guiseppe Occhialini of the University of Rome and Patrick Blackett of Cambridge University confirmed the discovery independently. Occhialini was working closely with the Cavendish Laboratory of Cambridge University. He had been working as an assistant under the famed Bruno Rossi and used their theoretical work on Einstein's mass/energy equation to tackle the cosmic ray studies.

Occhialini had theorized that the tracks of a cosmic ray appeared when an electron entered the chamber. Since the famous Einstein equation ($E=mc^2$) said that energy and mass are interconvertible, Occhialini theorized that the cosmic ray energy could convert to mass in the form of both positive and negative electrons. If correct, this transformation should show up in the chamber: the energy, for example, would convert to mass in the form of *pairs* of positrons and electrons. They, in turn, should leave the characteristic "tracks" found in cloud chambers. Occhialini decided to try to apply this theory.

To accomplish all of this, however, Occhialini had to conquer a formidable engineering difficulty—the inefficiency of the cloud chamber itself. A problem in studying cosmic rays was their unpredictability—you never knew when one might wander into the chamber. (Many kinds of rays easily penetrate ordinary materials, such as glass, steel, wood, iron, etc.) Occhialini's goal was to find a way to improve the timing of the cloud chamber, so that they might "catch" a cosmic ray when it entered the chamber.

Occhialini suggested placing Geiger counters just outside the chamber to tell them when a cosmic ray was entering the chamber proper. When a ray hit, the electrical impulse generated would automatically activate or "turn on" the chamber, so that it was prepared to photograph the ionized tracks caused by the ray (which tend to linger for some time after the cosmic ray itself has passed).

Using the counter-controlled cloud chamber, the camera was activated when an electronic pulse triggered the mechanism and photographed the incoming particle.

By November 1932, Occhialini, ably assisted by Blackett and still using the magnetic field, had photographed almost 1000 cosmic rays. The photos showed a virtual barrage of cosmic rays, some curving to the left and some to the right in the chamber's magnetic field, depending on whether they were positively or negatively charged—all precisely as his theory suggested. From that point on, the interest in cosmic ray studies swelled steadily. Later, Anderson even resumed his work on cosmic rays which, in turn, soon led to the discovery of the mu-meson.

Thus, in 1933, Occhialini and his associate Blackett were able to report "the same remarkable conclusion as Anderson," namely, that corresponding to electrons were positrons. Soon, this annihilated any lingering skepticism about Dirac's theory of the electron. So it was that, in the Solvay conference in October 1933 in Brussels, Dirac, with Anderson's and Occhialini's discovery in his pocket, could speak confidently in defense of his general approach to electron theory and quantum mechanics.

PARTICLES AND ANTIPARTICLES

Dirac's work was not the end of the story, merely the first installment. Once the positron had revealed itself, scientists began to suspect that Anderson may have found but a single instance of a universal phenomenon: for *every* particle, there might well be an antiparticle—a particle identical in mass but of opposite charge. Many also suspected that not every particle would follow the logic of the positron; it would not be possible to see all such antiparticles as "holes" in a "sea" of some kind. That, in turn, led to even more ineffable speculations about parallel universes of antimatter. (Bumping into your "twin" car in the antiuniverse would be catastrophic, however, since matter and antimatter annihilate one another.)

The next important development came in 1934, with Victor Weisskopf, an important pioneer in the early history of particle physics, now at MIT, and Wolfgang Pauli. Pauli, a German refugee physicist best known for his famous "exclusion principle," which stated that no two electrons could occupy the same place (quantum state) when in orbit around the nucleus of an atom, began to realize that *all* particles could have antiparticles. The question of how particles can have antiparticles is related to the principles governing the conversion of energy to matter, which say that particles and antiparticles must be created in equal numbers.

OUTLAW ELECTRONS: HOT ON THE TRAIL OF MUONS

But the work of that year did not make everyone happy. Anderson and others at Caltech realized that some electrons were not behaving as good electrons were expected to. The collisions involving electrons were not casting off enough energy. Both Hans Bethe and Walter Heitler realized that the calculated and expected *theoretical* emission of energy of some of Anderson's "electrons" was far too great to square with Anderson's experimental findings, which were much less than the expected, theoretical findings.

For a while, Anderson called his misbehaving electrons "green" electrons, while ordinary ones were "red." (Of course, fiddling with semantics didn't help. Possibly he just wanted to have a little fun *en passant*.)

ACCUSE QUANTUM MECHANICS OF CAUSING THE TROUBLE

So the problem remained. Again, Bethe and Heitler tried to come to the rescue with a bizarre notion, one that struck most as a desperate and ad hoc try—an explanation scarcely better than Anderson's. The theory was as follows:

> . . . . One should not expect that ordinary quantum mechanics which treats the electron as a point-charge [a negative charge but with no dimensions in space at all: literally a mathematical point]

> could hold under those conditions [i.e., within the unfathomably
> small distances of the atomic nucleus]. . . . It is very interesting that
> the energy loss of the fast electrons . . . provides the first instances in
> which quantum mechanics apparently breaks down for a phenome-
> non outside the nucleus. . . . We believe that the radiation of fast
> electrons will be tests for any quantum-electrodynamics to be con-
> structed.[13]

Give up quantum mechanics? The physics community would
have none of it. Fortunately, more sober minds soon realized that
Anderson's high-energy "green" electrons were in reality yet
another new particle dubbed the muon. This particle was 200
times as massive as an electron. And, since everyone knew that
according to Einstein's famous $E = mc^2$ any mass could be converted into energy, the massive muons explained the great excess
energy of the "green" electrons.

ANDERSON LATER IN LIFE

Later, Anderson would become active in the Office of Scientific Research and Development (OSRD). In a letter from the
Caltech archives, dated March 15, 1943, to William Lacey of the
OSRD from an unidentified person, reference is made to Anderson, as being a member of a committee designed to coordinate the
activities of Caltech with the U.S. Navy:

> Yesterday Dr. Millikan wired you requesting you to attend a confer-
> ence . . . at the New Statler Hotel in Washington. . . . There has been
> a committee composed of Sorenson, Chairman, . . . Carl Anderson,
> Ward . . . to consider the correlation on our activities with the navy
> program. . . .

Yet Anderson was hardly mutating into an administrative func-
tionary. His intellect, never dormant, was continually generating
new ideas and new lines of research.

Besides the Nobel Prize, Anderson received an honorary

doctorate from Colgate University, the Cresson Medal, and the gold medal of the City of New York.

Anderson died in January 1991 at the age of 85, but his work forever changed physics. With his contributions, the new era of "antimatter" physics took shape. Just as an antielectron or "positron" corresponded to the electron, physicists now believe that many particles have their own "antiparticles." In fact, there is now speculation about entire parallel universes of matter and antimatter—the great nexus between science fiction and science fact.

CHAPTER 13

Hans Bethe and Energy in the Sun

Hans Bethe, one of the world's preeminent astrophysicists for over 50 years, was born in Strasbourg in 1906. When he was barely 21 years old, he was already considered one of the finest students of Arnold Sommerfeld, easily ranking with his other students Peter Debye and Werner Heisenberg. He studied physics at the universities of Munich and Frankfurt and did his graduate work at Munich.

His first teaching post was at the University of Frankfurt. In 1932, the year Nazism was beginning to infect the world, Bethe was at Tübingen. In the German presidential elections that year, Hitler finished just behind Hindenburg and refused the latter's offer to become the vice-chancellor. In the Reichstag elections, the Nazis captured an overwhelming majority of the seats—far ahead of the socialists. The handwriting was clear enough and Bethe, who was Jewish, left for England and the University of Manchester in 1933. England was a logical choice, since great scientific work was going on there at the time. For example, English physicist James Chadwick discovered the neutron the previous year. Equally great breakthroughs were occurring in the United States. Working under Millikan, Carl Anderson discovered the positron at Caltech in 1932 and Irving Langmuir would capture the Nobel Prize in chemistry that year. Also, Harold Urey had just discovered the "heavy" isotope (a variation of an element where the nucleus has the same number of protons but a different mass than the standard form of the element) of hydrogen, or "deuterium," the same year.

Since the Manchester job was a one-year replacement position only, Bethe, in 1935, joined the Cornell University faculty as professor of theoretical physics, where he remains today. During WWII he worked on the Manhattan Project at the University of Chicago. After the war, he further explained and elaborated upon Bohr's theory of atomic structure which, in turn, led to his monumental investigations of energy production in stars.

By the time he landed in the United States during the early 1930s, Bethe was, of course, already well known, even though his finest work still lay ahead of him. Affable and outgoing, Bethe was always well liked by both students and faculty. In fact, during the 1960s, I often saw him in downtown Ithaca taking in a movie with his students.

ROBERT BACHER

One of Bethe's first tasks at Cornell was to produce, with Robert Bacher, former presidential science advisor and now retired from Caltech, a synopsis of the current work in nuclear physics. Bacher was a dedicated physicist and a former student of Samuel Goudsmit, the famed German refugee physicist who discovered that electrons spin on their axes while circling the nucleus.

The story of Bacher's coming to Cornell is sad testimony to the mania of the 1930s. In 1934, when he was being considered for the Cornell post, Goudsmit was asked by some minor Cornell scientists "whether Bacher was Jewish." It made an impression on Bacher and Bacher was thereafter interested in helping anyone who was Jewish.[1]

Working long hours at Rockefeller Hall, the physics building at Cornell under its legendary poor lighting, Bacher swam in mounds of paper, and took only occasional breaks for coffee at "The Straight" (Willard Straight Hall, the Student Union). It paid off. Bethe and Bacher, over the next two years, produced what is now called "Bethe's Bible"—certainly the best summary of nuclear physics up to that time.

Soon after that, scientists began working furiously in the prestigious new field of nuclear physics. Once a second-best stepchild to other areas of physics, it was rapidly beginning to dominate the pages of journals like *The Physical Review* and the *Journal of Astrophysics*.

THE FORCES WITHIN THE NUCLEUS

Since Bethe was a theoretician of the old German school, his addition to the ranks of American physics added yet another stellar name to the "theoretical" list, and drove another nail into the coffin housing the old "experimentalist" stereotype of the United States.

Bethe immediately attacked the problem of nuclear forces— the "glue" that cements the nuclear particles (protons and neutrons) together. To aid him in this research, he, in collaboration with his colleague, Birmingham physicist Rudolf Peierls, began constructing models of some of the lighter nuclei, such as hydrogen and helium.

The lighter nuclei were the logical starting place of course, since the combination of a single neutron and a lone proton is the simplest possible nucleus containing more than a single particle. Fully understanding its dynamics was absolutely fundamental for understanding all intranuclear interactions: more complicated nuclei could be viewed as multiples of this fundamental unit. The job was staggeringly complicated and seething with problems, but Bethe soon vanquished them.

One of the puzzles concerned the possible role of quantum mechanics in solving the structure of the nucleus. Still less than a decade old, quantum mechanics and its Schrödinger equation were powerful tools, already having proven their usefulness to theoreticians like Langmuir and Sommerfeld in understanding the phenomena of chemical bonding. Yet there was a difference—possibly a critical one. Chemistry, since it concerned the behavior of particles outside of the nucleus, did not necessarily constitute evidence that the new science would prove helpful in understanding what went on inside the nucleus.

In order to facilitate their work, Bethe and Peierls simply assumed that the proton–neutron complex could be described by a single law, and applied the Schrödinger equation, the main tool of quantum mechanics for analyzing the behavior of subatomic particles (electrons around a nucleus, for example). This famous equation describes the behavior of particles by predicting not a particular definite future behavior of the particle, but rather a series of possible behaviors. It also gives the probability that each possible behavior will, in fact, occur. According to their model, as long as the proton and neutron were at least 10^{-13} centimeters apart, they would not affect one another. For any distance less than that, quantum mechanical calculations showed that a very large, stable attractive force—the "rectangular well" as physicists picturesquely described it—would exist between them; that is, the proton and neutron would hold one another together comfortably so that the nucleus remained intact.

For a distance greater than 10^{-13} centimeters the proton and neutron would drift apart. (This so-called "two-body system" was merely an isotope of deuterium or "heavy hydrogen.") So, Bethe had not only cleared a gigantic hurdle in our understanding of the behavior of particles within the nucleus of the atom, but had proven clearly that quantum mechanics was indeed useful both outside and inside the nucleus.

THE ENERGY OF THE STARS

These insights made some theorists suspect that clues lay in Bethe's work which could improve our understanding of the mechanism of energy production in stars. Some things were already known: scientists knew, for example, that ordinary chemical reactions could not possibly generate the kind of energy seen in stars. For one thing, since no molecule can exist intact in the enormous temperatures seen inside stars, there simply would *be* no molecules to combine chemically with each other. Thus, ordinary chemical combustion cannot occur. (You can't light a match if there are no matches.)

Nor could such energy be explained by the gradual collapse of a star through gravitation—a star's being crushed to a fantastically dense state by its own gravitational pull (sort of like a building collapsing because the facades are too heavy). If the sun really were collapsing, it would have completely done so ages ago after, roughly, only 30,000,000 years of existence. However, scientists knew the sun was much older than that. Analogously, an extremely bright star such as Capella would have collapsed completely within about 300,000 years. Again, astrophysicists knew Capella was considerably older than that.

So the gravitational-collapse theory was out. Nor could the energy production be explained by the sun's picking up some amount of matter from outside. Any source of energy that could answer the fundamental question had to come from inside the star rather than from the surface, because there had to be an explanation for the extraordinarily high internal temperatures. Without these temperatures, the core of the star would not have enough energy to counterbalance its tendency to collapse by the force of its own gravity.

It was no surprise that scientists soon turned to Bethe's work for an explanation. Many had believed for some time that the generation of the energy seen in stars involved nuclear reactions of some sort, even though they knew little beyond that bare fact. But even when physicists suspected that nuclear reactions might be the answer, they realized that they did not understand either how such reactions began or what sustained them—until Bethe came along.

Sir Arthur Eddington, British astronomer and popularizer of science, believed as early as 1930 that stellar energy was in fact produced by nuclear reactions. As he said in his book, *The Internal Constitution of Stars*:

> It is now generally agreed that the main source of a star's energy is subatomic. There appears to be no escape from this conclusion; but since the hypothesis presents many difficulties when we study the details, it is incumbent upon us to examine carefully all alternatives. . . .
>
> In seeking a source of energy other than contraction [gravitational] the first question is whether the energy to be radiated in the future is hidden in the star or whether it is being picked up continuously from outside. Suggestions have been made that the impact of meteoric matter provides the heat or that there is some subtle radiation traversing space which the star picks up. Strong objections may be urged against these hypotheses individually; but it is necessary to consider them in detail because they have arisen through a misunderstanding of the nature of the problem. *No source of energy is of any avail unless it liberates energy in the deep interior of the star.* It is not enough to provide for the external radiation of the star. We must provide for the maintenance of the high internal temperature without which the star would collapse.

FUSION AS THE SOURCE OF ENERGY PRODUCTION

Eddington, of course, considered many types of energy-generating processes in his book. Energy can, for example, be produced by the disintegration of atomic nuclei as in ordinary radioactivity. However, this could not account for the enormous energy in the output of a star, nor for the enormously long lifetimes of stars. Any element that decayed radioactively would have decayed completely a long time ago. And if that were the principal means of energy production in our sun, our sun would have reached a cold, dead state eons ago and we wouldn't be here to wonder about it at all.

As Eddington rightly concluded, the only mechanism of energy production that could explain such facts is *fusion*—the joining together of smaller atoms to yield larger ones, with a consequent release of energy. Naturally, Bethe also knew, through Einstein's $E=mc^2$, that when any amount of mass is "destroyed" or converted into another form, an equivalent amount of energy is released. Thus, the "fusion" of two atoms might be the explanation he was seeking.

Bethe began by analyzing a set of nuclear interactions starting with hydrogen nuclei and ending with their fusion to form the helium nucleus. The series—now known as the proton–proton chain reaction—was subsequently studied and corroborated by many other scientists working independently. This is a series of thermonuclear reactions whereby hydrogen is converted to helium with an accompanying release of energy.

Still, even by 1938, Bethe himself was not convinced that this was the whole story behind stellar energy production. Based on the data then available, he concluded that this proton–proton mechanism could explain only part of the sun's energy production. So, he looked elsewhere to account for the "missing" mechanisms.

MISSING ENERGY FURNACES

Soon he found the kernel of one of his most ingenious and deservedly heralded contributions. It came to him after attending a conference devoted to this and similar issues, at the annual meeting of the American Physical Society in Washington. Allegedly, on the train ride home, he thought about all of the possible ways that lighter atomic nuclei could interact with the simplest hydrogen nucleus in a star's "furnace." He did this with elements starting from helium and going almost up to carbon. But he soon ran into the same kind of problem that had doomed the "radioactive decay" theory of energy production mentioned earlier. It quickly became clear to him that such processes, while they might account for the energy production of a star, would occur much too quickly to explain the long life of a star. It is a matter of simple observation and calculation that the conversion to the inert element helium occurs so rapidly that, were this the only mechanism of energy production, there would be no more stars "alive" today. (This is similar to why romantic couples wishing to pass a long evening by the fireplace have to fill it with wood rather than sheets of newspaper, which burn up immediately.)

THE CARBON CYCLE

The key lay in some peculiar properties of the element carbon. Carbon is different from other elements in that while it reacts similarly to the other atoms he'd considered, it does not convert to helium in one step. He had hit upon the "carbon cycle." With Bethe's "carbon cycle," although the conversion of hydrogen to helium occurred too quickly to explain the life of a larger star, the conversion of hydrogen to helium, with carbon acting as a catalyst, did *not*. (Again, this is like the need for wood, rather than newspaper, to keep the fireplace going for a long time.) To change simple elements to carbon, six time-consuming internuclear reactions occur, which are then, finally, converted into a completed helium nucleus.

The key lies in the enormous temperatures—up to 20 million degrees—in the center, a figure easily calculated merely by extrapolating from the surface temperature. Earlier, German physicist Weizsacker had also proposed on his own that the course of energy in stars was "fusion," or the joining of two hydrogen nuclei to form helium. However, he did not realize that there was the problem with the rate of production of helium as discussed above. Bethe knew that thermonuclear (basically a series of fusion reactions energetic enough to keep the reactions going) reactions are the chief source of energy in the solar "furnace."

Bethe subsequently realized that the process consists of an entire sequence or "chain" of thermonuclear reactions.[2] As Bethe explained it, the chain is a closed circle: it returns to the starting point after six cycles, much as a runner would if he was running around a circular track. The players in the game are carbon nuclei, nitrogen nuclei, and the protons that smash into them. If carbon collides with a proton, then an isotope (a version of an element with the same number of protons, but different numbers of neutrons) of nitrogen is produced. Also and at the same time, considerable amounts of energy are freed as gamma rays. However, the nitrogen isotope, being unstable, next transmutes itself into an unstable form of carbon by ejecting a positron (an ordinary electron, but with a positive charge). The isotope of carbon is again hit by a proton and turned into a common and stable form of nitrogen, again with an accompanying release of high-energy gamma rays.

The stable nitrogen isotope next collides once again with a proton, yielding more energy and an unstable form of oxygen. The oxygen, in turn, quickly transmutes into a *stable* isotope of nitrogen by emitting a positron.

Ultimately, the stable isotope of nitrogen is again struck by a proton and splits in half. One half is an ordinary carbon nucleus; the other is an alpha particle—a helium nucleus. The chain can be repeated virtually endlessly, since it continually returns to exactly the same element that started it—carbon. And since one of the final products is also an alpha particle, or helium nucleus, in effect the carbon cycle transmutes hydrogen nuclei (the protons) into helium. In other words, the entire cycle is essentially a fusion reaction, though it is accomplished in a very roundabout way. (Producing helium nuclei in this way can be compared to changing money: you can convert dollars into francs directly or—in the spirit of Bethe's carbon cycle—you can change dollars into Italian lire, lire into German marks, marks into Japanese yen, finally changing yen into francs.)

In fact, Bethe proved that the theoretical amount of energy that should be liberated by this theory matched the observed energy output of the sun, based on data then available. Bethe further noted that, at the then-accepted temperatures of 20 million degrees[3] in the center of the sun, the process would take about 5 million years. The extra steps and thus the extra time needed to complete this cycle squared with estimates about the lifetime of other stars as well. Bethe finally published these results in 1938 in the prestigious journal *The Physical Review*.[4] (His papers around this period give the impression that the carbon cycle was the chief energy mechanism in virtually all stars, no matter their size.)

PROBLEMS IN THE THEORY

Still, problems remained: How could a proton be absorbed by the other nuclei in Bethe's scheme? Since a proton was positively charged, any other nucleus would tend to repel it by very powerful electrostatic forces (much as the positive ends of two ordinary household bar magnets will repel each other). If this theory were truly the answer, how could such extremely strong electrostatic repulsions be overcome?

One possible explanation hinged on the fact that such repulsion could actually be overcome when the proton and the positive nuclei smashed into each other with enough force. But physicists knew that even inside a star, where protons and nuclei are both moving with tremendous velocities, they weren't moving quite fast *enough*—even with temperatures in the millions of degrees. Of course, no two protons move at exactly the same speed. Therefore, at times some might be moving with the required velocity. Still, the mathematics just didn't work out. Not enough of them would be traveling at the required speeds to explain such fantastic energy production.

QUANTUM MECHANICS SAVES THE DAY

But once again, quantum mechanics came to the rescue. The problem was that protons—as particles—could not possibly have enough energy to overcome the electrostatic repulsion between them and positively charged nuclei. But one of Schrödinger's and de Broglie's triumphs was to show that particles such as protons had *wave* characteristics as well as particle characteristics. Bethe had an idea: it was possible that as particles, the energies would not be enough, but as *waves*, they might be. He performed some probability calculations based on the assumption that the *wave* properties of protons would allow them to collide with the positively charged nuclei as the carbon cycle required—even when, as particles, the protons would not have enough energy to overcome electrostatic repulsions. (In a sense, Bethe is saying that waves are more energetic than particles.) With his probability calculations, he found that knowing the rate at which the carbon cycle occurred, and considering the wave properties of protons, the speed *was* great enough to explain the temperatures at the center of stars similar to the sun.

Some of Bethe's figures naturally had to be adjusted with the advent of more refined measuring techniques. Astrophysicists now believe that there are basically two sources of energy production: the carbon cycle and the so-called "proton–proton" chain, where a nucleus of hydrogen (which is really just a proton) is converted directly to helium without carbon playing any role. They further believed that Bethe's carbon–carbon cycle is the chief energy source *only* for stars larger than the sun. Bethe himself later further developed the proton–proton chain idea and showed it was the chief source of energy production in both the sun and stars smaller than it, though both go on simultaneously. In short, for stars like the sun or smaller, the proton–proton mechanism is dominant, while for stars larger and brighter than the sun, the carbon cycle is dominant.

THE NEAR DEMISE OF THE AUFBAU THEORY

Based on this work, Bethe also believed he had refuted what had been known as the "Aufbau" theory: this idea maintained that all of the heavier elements are built in successive steps, merely by the addition of protons. The problem, as Bethe saw it, was that there was no element with an atomic weight of five. This, in turn, indicated that there would be an unbridgeable gap between helium and lithium. The capture of a proton—the heart of the Aufbau idea—could not explain this transition.

However, subsequent research showed that when enough hydrogen has been changed into helium, the helium undergoes a sudden and powerful contraction, producing temperatures of nearly 100 million degrees. At that point another kind of transmutation occurs, where three alpha particles join to produce ordinary carbon, thus bypassing the lithium gap. After that, with increasing temperatures, protons can be added successively to produce increasingly heavier elements as the Aufbau theory predicted.

LATER WORK

Along with physicists R. A. Alpher and George Gamow of George Washington University, Bethe did considerable work that helped explain the huge amounts of helium in the universe and that also supported the big bang theory of the creation of the cosmos. A by-product of their theory was the prediction that there should be a wall of microwave radiation filling all of space—"left over" until today from a very hot early period in the history of the universe. (Microwaves are similar to ordinary light waves, but lying between the radio and infrared regions of the electromagnetic spectrum.)

For years after their work, this prediction remained unproven. But one day, Bell Labs physicist Arno Penzias and his colleague Robert Wilson were testing a microwave detector when they suddenly found it was producing a lot more noise than they had anticipated. Perhaps most importantly, the noise level did not change no matter how they changed the direction of the detector. That, in turn, indicated that it was coming from outside the atmosphere and outside the solar system, thus confirming one of the major predictions of the theoretical work by Bethe, Gamow, and Alpher on the big bang theory. Penzias had found the microwave radiation predicted by Bethe and Gamow, winning Penzias the Nobel Prize.

It was due largely to the Gamow/Alpher/Bethe theory, and Penzias's confirmation of their theories that the big bang concept finally eclipsed its rival, the "steady state" theory (which held that the universe, though expanding, remains at the same density— the latter claim distinguishing it from the big bang theory). Primarily as a result of Bethe's work on solar energy production, he was awarded the Nobel Prize in physics in 1967.

BETHE AND NEUTRINOS FROM THE SUN

Most recently, Bethe has endorsed an odd notion formulated by S. P. Mikheyev and A. Y. Smirnov of the University of Moscow. A discovery by American astrophysicist Ray Davis of the Brookhaven Laboratory prompted the theory. Davis knew that neutrinos, or subatomic particles with no electrical charge, come in a variety of forms, including so-called "tau" neutrinos, or neutrinos that appear simultaneously with the production of "tau" particles. The latter is a subatomic particle many times more massive than an electron.

Still another kind of neutrino is associated with the production of "muons," or subatomic particles similar to tau particles, but far more massive. There are also "electron" neutrinos, or neutrinos associated with the production of ordinary electrons. Davis also knew, as did all physicists, that current particle theory, as well as Bethe's views about how energy is produced in the sun, predicted that a certain number of electron neutrinos should arrive at the

Earth from the sun in a given amount of time. To detect them, Davis built a huge tank of carbon tetrachloride, or ordinary cleaning fluid, about a mile underground, to prevent cosmic rays from disturbing the carbon tetrachloride container. When hit with neutrinos, the chlorine in the carbon tetrachloride should change into the inert gas argon. (Neutrinos do not affect the Earth in any other important ways.) The amount of chlorine converted to argon would then allow one to calculate the number of neutrinos entering the chlorine. Amazingly, however, Davis found that only about one-third of the expected neutrinos from the sun were actually arriving on the Earth.

The importance of the neutrino theory is that such neutrino production would have been evidence for Bethe's theory about solar energy production, since the liberation of neutrinos is predicted by his theories as described earlier. But since far fewer than the predicted number were found, something was wrong. One possibility was that energy production on the sun was two-thirds less than anyone had ever suspected, which seemed almost too bizarre to contemplate.

The other possibility was the Mikheyev–Smirnov theory. The basic idea of the Mikheyev–Smirnov idea is that neutrinos— infinitesimally small, neutral, energy-carrying particles—have, in line with quantum mechanics, both wave and particle characteristics. They sometimes behave as particles and sometimes as waves. Now the Soviets believed that ordinary changes from one type of neutrino to another would be affected by the matter they happened to slam into while traveling through the interior of our sun. Such collisions would raise the energy of the neutrino and increase its chance of changing into another type of neutrino. So the explanation of Davis's shortfall is then obvious: the reason he found only a third of the expected number of electron neutrinos is that the others must have mutated into other kinds of neutrinos on their way to us, perhaps into "tau" neutrinos. The point then is that these *other* kinds of neutrinos simply have no effect on the carbon tetrachloride and do not change the chlorine into argon as the "electron" type of neutrino would have done. Although the theory is still highly speculative, it did receive a tremendous boost when Bethe endorsed it—further evidence of the enormous prestige he still holds in scientific circles.

Now retired from Cornell, he pursues his theoretical work in astrophysics and can still be seen occasionally at the movies in downtown Ithaca.

CHAPTER 14

The Manhattan Project: Witness to the Atomic Age

In 1939, German physicist Lise Meitner and German chemist Otto Hahn proved for the first time that a sufficiently powerful ray of neutrons could split the atom. This, in turn, led physicists to begin to speculate about the feasibility of a new and terrible weapon—the atom bomb. Scientists knew that a powerful bomb might be possible, mainly because of Danish physicist Niels Bohr's "liquid-drop" model of the atom. In Bohr's model, published in 1939, the nucleus is a droplet made up of protons and neutrons. Bohr believed that because of the electrostatic repulsions between positively charged protons in the nucleus, tremendous energies were required to hold the nucleus together. Thus, based on purely theoretical considerations, they speculated that this energy, if it could be released, would constitute a tremendously powerful weapon.[1]

But there were different problems to consider aside from the scientific ones. Physicists recognized that the Germans would be thinking along the same lines. After weeks of discussion with other scientists, American physicists realized that if the United States was to build an atom bomb first, the government had to be informed of their intentions. George Pegram of Columbia University made the first contacts with the White House regarding the possibility of a atom bomb.

ROOSEVELT CONVINCED; MANHATTAN PROJECT FOUNDED

On December 6, 1941, the Office of Scientific Research and Development (OSRD) took charge. Involved at this point were James Conant, L. J. Briggs, Harold Urey, E. O. Lawrence, Arthur Compton, E. U. Condon, and many other distinguished physicists. The Manhattan Project was now at full throttle and responsibility for it was turned over to General Leslie Groves. Although Bush remained as general overseer of the project, Bush realized that the scope of the project needed the cooperation of both industrial and civilian scientists of all types. He realized that orchestrating all of this was beyond the administrative competence of the OSRD. Thus, he arranged for all of the engineering, materials acquisition, and scientific research to be turned over to Groves.

Groves had had considerable experience as manager of the building of the Pentagon and was renowned for his leadership abilities as well as his considerable technological prowess. Also, Groves knew his limits and delegated the scientific work to others. He turned over, for example, the problem of separating the usable uranium to chemist Harold Urey of Columbia and the problem of studying chain reactions to Arthur Compton of the University of Chicago.

THE PILE

At the beginning of the Manhattan Project, the center of activity quickly became the University of Chicago. That is where the government was building the first atomic pile and where the first atomic chain reaction would hopefully be created. By late December 1942, early work on the chain reaction began. Under the leadership of the great Italian physicist Enrico Fermi, the physicists on the Manhattan Project had obtained enormous amounts of uranium and had started developing techniques for controlling, starting, and measuring the force of nuclear reactions.[2]

The tension was unfathomably high. Everybody knew the potential dangers of radioactivity as well as the possibility of a nuclear reaction escalating out of control. Inside the pile itself were both safety and control rods. With them scientists could adjust the rate of the reaction either by withdrawing or by further inserting them. A paradigmatic balancing act: fully inserted, they would absorb too many neutrons for a chain reaction to proceed. Fully withdrawn, the pile would get out of control and produce an atomic blast.

December 2 was the eve of the atomic age. Everyone assembled at 8:30 on that Wednesday morning in the squash court of the University of Chicago. Zinn and Anderson continued making measurements of what was going on inside of the pile—convinced that a reaction could be made self-perpetuating, meaning that it could continue on its own, without any interference from scientists.

THE ITALIAN NAVIGATOR LANDS IN THE NEW WORLD

The "suicide squad" prepared themselves. Positioned dangerously close (though "safe" scarcely has any meaning in such circumstances) to the pile on a platform above the pile, they were the men at greatest risk, for they had the onerous task of destroying the pile and trying to halt the reaction if anything went wrong. This squad consisted of engineering safety specialists Nyeter, A. C. Graves, and Harold Lichtenberger. If catastrophe loomed, they would douse the pile with a cadmium salt liquid to halt the reaction.

Once the scientists accomplished the first successful chain reaction, Arthur Compton called James Conant. Using the wartime code, he merely said, "Jim, you'll be interested to know that the Italian navigator has just landed in the new world," meaning of course that Fermi's group of physicists had artificially produced and controlled atomic fission for the first time in history.

THE BUILDING OF THE BOMB

The remaining task was to build the atom bomb itself. General Groves then further delegated responsibility for various stages of the work. Lawrence of the University of California would handle the plutonium work, Urey would further improve gaseous diffusion (a method of separating U-235 from U-238), and the study of fission would be supervised by Compton. Ultimately, two bombs would be built, one of plutonium, the other of uranium.

ON SEPARATING THE ISOTOPES OF URANIUM

One of the earliest problems was obtaining the uranium for the bomb. That meant separating the useless U-238 from the fissionable U-235. But how could this be done? Of the many ways of separating the two isotopes of uranium, two were selected, for the most part because they were, despite many technical difficulties, able to separate the isotopes fastest and most efficiently. One was the electromagnetic process developed at Berkeley by Lawrence, and the other was the "diffusion" method worked out by Urey at Columbia. With the electromagnetic system, uranium atoms were placed in a magnetic field and the smaller, fissionable U-235 atoms were collected in a box. The system depended on the fact that the trajectory of U-235 atoms in a magnetic field is slightly different from that of U-238.

UREY AND GAS DIFFUSION SEPARATION

Harold Clayton Urey was born in 1893 on a rural Indiana farm. He graduated with a bachelor's degree in zoology from Montana State University in 1917 but, under the influence of British biologist Archie Bray, switched to chemistry. He entered the University of California at Berkeley, studying under G. N. Lewis, perhaps the greatest American chemist of the day. In 1923, he received his doctorate under Lewis. His great work was the discovery of a heavier form of hydrogen, for which he won the Nobel Prize in 1934.

Urey was perhaps one of the kindest of all great scientists. He is justly renowned for coming to the aid of Nobel Prize-winning physicist Maria Mayer. For many years, Mayer had had to live ". . . without a country," as described by her husband Joe Mayer in a June 12, 1940 letter to Linus Pauling. Because her husband was already on the faculty, the University of Chicago would give her only a nonpaying teaching post. Urey, to aid her, often arranged paid lectures. That, in turn, led to her also working on the Manhattan Project. He arranged a half-time position for her in his work on separating the isotopes of uranium.

Urey's method, supported by British scientific opinion, relied on the fact that gaseous U-235 would diffuse through small holes at a rate slightly different from U-238.[3] But both methods of separation were immensely costly and difficult. Thus, there was still very little progress in acquiring U-235, even well into 1944.

General Groves was desperate; he thus pinned his hopes on still a *third* unknown and enormously speculative technique for separating uranium isotopes—thermal diffusion. In this technique, the separation depended on the fact that atoms of differing weights tend to separate from one another between hot and cold regions. But since this too proved only guardedly effective, Groves decided to continue research on all three methods simultaneously at the Oak Ridge National Laboratory, newly created and equipped specifically for atom bomb work.

GLENN SEABORG: PLUTONIUM AND THE
TRANSURANIUM ELEMENTS

Then a miracle occurred. In early 1941, researcher Glenn Seaborg, a youthful physical chemist at the University of California at Berkeley, resumed some of Enrico Fermi's earlier research on transuranium elements (i.e., elements heavier than uranium). Working with his colleague Edwin McMillan at Berkeley, the pair bombarded uranium nuclei with neutrons. Although theoretically predictable, Seaborg was nonetheless delighted to find that some of the nuclei did not split when hit by a neutron. Instead, they absorbed it and underwent beta decay—the spontaneous ejection of electrons—as predicted by Fermi. Seaborg further realized that, if left alone, the transmutation would eventually turn into still *another* new element called neptunium and then into plutonium. At one stroke, the supply problem was solved. Plutonium, as fissionable as uranium, could be manufactured in the laboratory at manageable costs.

By 1942, scientists determined that a powerful explosion could be produced using less than 100 pounds of pure U-235 or plutonium. Such an explosion would equal the destructive energy of 20,000 tons of TNT. Even so, the estimated cost was formidable: over $1 billion, not to mention the need for several more years of work.

During this time, Germany would be conducting their heavy water experiments. Fortunately, the choice of heavy water was a mistake by German physicist Walter Bothe, who, because of a poor experiment, wrongly concluded that graphite would not do well as it absorbed too many neutrons. The point was that learning how to construct a *controlled* nuclear reaction was the first step in building an actual bomb. Heavy water was a good absorber of neutrons and would therefore control a nuclear reaction very well. But the only plant capable of making heavy water in Europe was in the city of Vemork in northern Norway. Knowing this, the Allied forces destroyed the plant and the Germans were never able to get enough heavy water for their experiments. And having already ruled out graphite, their work proceeded too slowly to constitute a real threat in the "race" for the atom bomb.

OPPENHEIMER'S ROLE: DELIVERY PROBLEMS

The great orchestrator of the entire project was the brilliant theoretical physicist J. Robert Oppenheimer. Oppenheimer was among the greatest early theorists in American physics. He graduated from Harvard in three years,[4] summa cum laude, in 1925—the great year of quantum mechanics. After a not terribly productive year at the Cavendish Laboratory in England, he accepted an offer to work at the University of Göttingen with the fabled physicist Max Born. It is clear from his correspondence that, even at this early period in his life, he was interested in problems relating to nuclear physics. As he wrote to his old Harvard physics professor Edwin Kemble on November 27, 1926,

> Dear Dr. Kemble: Many thanks for your kind letter. . . . This term I am spending at Gottingen. . . . [A] problem on which Prof. Born and I are working is the law of deflection of, say, an [alpha]-particle by a nucleus. We have not made very much progress with this, but I think we shall soon have it. . . .

In 1930, among his other accomplishments, he proved that the "holes" Dirac had described in his epochal studies of the electron had to be positive particles of about the same mass as an electron. Later, that theory was confirmed by Carl Anderson's discovery of the positron in 1932.[5] When the announcement of the splitting of the uranium nucleus fell like the hammer of Thor on the world community, Nobel laureate Glenn Seaborg described its effect on Oppenheimer:

> I remember . . . a seminar in January, 1939 when new results . . . on the splitting of uranium with neutrons were excitedly discussed; I do not recall ever seeing Oppenheimer so stimulated and so full of ideas.[6]

On the surface he appeared much like a typical insurance sales-
man with his crew cut and conservative style of dress. But, ever
the mystic, he would often quote from the sacred Hindu text, the
Bhagavadgita.

Although Los Alamos really did not begin its scientific and
engineering research until April 1943, Oppenheimer had been
busy well before that. In the spring of 1942 the OSRD had named
him as director of the physics research needed for the bomb.
Arthur Compton's recommendation sold the government on Oppenheimer. Not only was he one of the world's greatest physicists,
but, according to Compton, he ". . . was one of the very few
American-born men who had the professional competence, and
he had demonstrated a certain firmness of character."[7] Soon, the
39-year-old Oppenheimer was directly in charge of the site in Los
Alamos where scientists designed and built the first bombs.
Negotiating directly with General Groves, Oppenheimer, Groves
and their aides decided on the remote area of Los Alamos for both
security and safety reasons.

Everything came together on July 16, 1945.

The surrounding throng of scientists and dignitaries waited
at a distance of several miles. In the doorjamb of the south bunker
(a small concrete-reinforced building used to observe the blast at
close range) stood Oppenheimer, tense, nervous, and perspiring
profusely. On top of a jeep was Feynman, playing his bongo
drums. The bomb exploded at 5:30 AM on July 16 on the Ala-
magordo Air Base, over 120 miles south of Albuquerque, New
Mexico. First came a brilliant burst of light, soon followed by the
tremendous blast of the explosion, followed again by a powerful
heat wave. Then came the now-familiar giant mushroom cloud
soaring nearly eight miles into the sky. Not only was the steel
tower vaporized, but the desert sand had turned to glass for over
1000 feet in every direction. Just 20 seconds after detonation, the
mushroom cloud was clearly visible far above the town of San
Antonio and over much of the desert.

In scientific and military terms, it was a grand success. At the
same time, a line from the *Bhagavad Gita* went through Op-

penheimer's mind: "I am Death, The shatterer of worlds." Armageddon was at hand. Witnesses had seen the dawn of the atomic age.

ARMAGEDDON

For Japan, the end was near. On August 6, 1945, three B-29s left Tinian Island in the Marianas and headed for Hiroshima. In the first plane, the *Enola Gay*, rested "Little Boy"—the first uranium atom bomb used in wartime. At 7:30 in the morning, Captain Bill Parsons, chief of the ordnance division, finished arming Little Boy and gave the OK for it to be loaded onto the plane. A few seconds after 8:50 in the morning, Hiroshima was in ruins, shrouded in a dust cloud that climbed six miles into the stratosphere.

The White House, that same day, issued a terse news report: "Sixteen hours ago an American airplane dropped one bomb on Hiroshima. . . . That bomb had more power than 20,000 tons of TNT. . . . It is a harnessing of the basic power of the universe." Three days after that, Nagasaki was obliterated as well by a plutonium bomb named Fat Man.

THE AFTERMATH

The reaction in America was mixed. While there were some strong public condemnations of the use of the bomb, most Americans believed as Truman did, that the bomb saved many American lives and ended the war far more quickly than any other policy, such as a direct land invasion of Japan, or a mere "demonstration" to Japan of what they would face if they did not surrender.

Yet even amid the public jubilation over the early end of the war, the public worried that, in splitting the atom, science had now ventured into territory where only the Almighty should go. Also, almost everyone, including pacifists and hawks in both the public and scientific community, believed that a strong, sustained program of national security was now needed. Pearl Harbor was still fresh in America's collective memory.

Then, too, American public opinion was deeply concerned over certain ominous happenings. Many were disturbed, for example, that German physicist and spy Klaus Fuchs had given atomic secrets to the Soviets who, in fact, had begun their atomic research in 1943 and had a bomb by 1949. The public quickly began to fear that soon, many nations would have the atom bomb and that mass destruction would be the inevitable consequence of man's scientific folly.

POSTWAR SCIENTIFIC RESEARCH

Scientists were also worried about the progress and control of scientific research in the postwar period, the ever-watchful Vannevar Bush among them. By the spring of 1944, he cautioned the OSRD advisory council that a plan was needed for peacetime research, since the OSRD automatically vanished at the war's end. One immediate consequence of Bush's warning was that in a letter of November 9, 1944, Secretary of War Stimson and U.S. Navy Admiral Forrestal asked the National Academy of Sciences to create a Research Board for National Security, under the guidance of Frank Jewett, then head of the Academy. Jewett appointed Karl T. Compton to chair it and the Navy gave it their full blessing.[8] Scarcely had the board been created, however, when President Roosevelt, in March 1945, ordered government research funds slated for the Board to be frozen until the chaotic state of postwar research could be clarified. Although the Board had accomplished some things in its short life, such as starting people thinking about the need for military research in America after the war, it was suddenly doomed. Despite the best efforts of Bush, Compton, and other scientists, no one could convince Roosevelt that the Board served any useful purpose. So it was that, in a letter dated

October 18, 1945, Forrestal and Patterson, acting under Roosevelt's instructions, obliterated the Board.

But the scientists heavily involved in the ill-fated Board persevered. Many were suspicious of Roosevelt, believing it was his plan to control postwar scientific research himself. Tragically, Roosevelt died in Georgia on April 12, 1945, before that question could be resolved.

Then too, the public, for the most part and despite the controversy surrounding the use of the atom bomb, supported the peacetime uses of atomic power along with controlled and carefully monitored sharing of nuclear secrets with other nations, especially the Soviets. Doubtless the strongest boost in public approval for the peacetime uses of atomic power emerged during the threat to the U.S. oil supply in the 1970s brought about by the most powerful oil-producing Arab nations. Only the catastrophes at Three Mile Island and Chernobyl have dampened this enthusiasm in recent years.

CHAPTER 15

The Myth of Symmetry:
Yang and Lee

> How would you like to live in a Looking-
> glass, Kitty? I wonder if they'd give you
> milk in there? Perhaps Looking-glass milk
> isn't good to drink.   —Alice

Lewis Carroll was right, even if he didn't know why. In fact, looking-glass milk is poisonous. The importance of parity (symmetry) violations is best understood against a backdrop of a venerable idea in physics. One of the most cherished visions in nature has always been that of a universe that is stable, orderly, and predictable. Briefly, the underlying idea of symmetry is merely that—the universe does and always will display a high degree of symmetry. The beautiful symmetry of a snowflake is one example, as is the "radial" symmetry of a starfish and the elegantly beautiful symmetry of the double-helical DNA molecule. Yet the world had to await the great physicists C. N. Yang and T.-D. Lee for physical proof of Carroll's ideas. As Yang himself described symmetry,

> One of the symmetry principles—the symmetry between the left and
> the right—is as old as human civilization. The question [of] whether
> nature exhibits such symmetry was debated at length by philoso-

phers in the past. Of course, in daily life, left and right are quite distinct from each other. Our hearts, for example, are always on our left sides. The language that people use, both in the Orient and the Occident, carries even a connotation that right is good and left is evil. However, the laws of physics have always shown complete symmetry between the left and the right. . . .[1]

Thus, a molecule that twists to the left should work the same way as one that twists to the right.

T. D. LEE—EARLY LIFE

Tsung-Dao Lee was born on November 24, 1928, the son of an impoverished Chinese merchant, Tsing-Kong Lee. The time of his birth seemed propitious as Chiang Kai-Shek was elected that year as president of China amid a plethora of promises of a brave new world for the Chinese people. But that never materialized and Lee grew up in the squalor and despair rampant in China in the wake of the worldwide depression heralded by the great crash of 1929. He attended the Kiangsi Middle School in Kanchow, China, graduating in 1943. He then began his college work at the National Checkiang University, where his genius in physics and mathematics immediately became obvious, but had to abandon his studies because of the assault by Japan.

At the war's end, he traveled to Kunming, Yunnan, attending the National Southwest Associated University, where he met his equally brilliant collaborator C. N. Yang for the first time. Soon, whispers of brilliant work seeped from the university to the highest levels of the Chinese government. Realizing Lee's genius, the Chinese government offered him a scholarship, which he used to attend the University of Chicago in 1946 to study with the great Italian physicist Enrico Fermi. Unlike his friend Yang, Lee's early interests were not in particle physics, but in astrophysics. He studied the evolution of stars at Chicago, particularly the properties of supercondensed matter. His dissertation was titled "Hydrogen Content of White Dwarfs," for which he received his Ph.D. in

1950. But the grandeur of the macrouniverse didn't enthrall him for long. He soon turned to the mysteries of the microworld of subatomic physics. His research quickly moved from the stupendously large cosmos of the galaxies to the vanishingly small world of the atom.

Destiny, in turn, took him to the University of California at Berkeley where he took the post of research associate. The following year, in 1951, he left California for the Institute for Advanced Study in Princeton, and remained there until 1953. At Princeton he began his serious collaboration with Yang, working on problems in particle physics and meson decay. While at the Institute he published a wide variety of papers on issues ranging from elementary particle physics to statistical mechanics. Most of these, while useful, were rather minor, routine additions to current atomic theory.

But he was scarcely interested in routine work. More and more, the paradoxes and odd behaviors of particles at the subatomic level fascinated him. Most importantly, it was at Princeton where the behavior of mesons began to bother him. Something was not right. Two mesons that appeared to be in every other way identical, nevertheless decayed in different ways. He had no clear ideas about how to solve this paradox.

Believing that he needed a change of atmosphere for inspiration, he left Princeton in 1953 to take the post of assistant professor of physics at Columbia University. He was appointed full professor in 1956; at the age of 29, he was the youngest full professor at Columbia. But his genius required complementation by Yang's. Soon, he was working intensely with Yang once more. Eventually, their joint efforts resulted in a 1956 paper published in *The Physical Review*. For the first time in the history of science, the principle of the conservation of parity was questioned by Yang and Lee. Both began to suspect that the laws of physics do in fact distinguish between right and left, contrary to the wisdom of the centuries. In the same paper, they suggested how experimental confirmation for their view could be attained. In 1957 they captured the Nobel Prize for their work.

After receiving the prize, it dawned on him that he could plumb the ultimate resources of his mind only with Yang. So he returned to Princeton to continue their collaboration on particle physics. Eventually, however, he received an offer that couldn't be refused, Yang notwithstanding. He was offered, and accepted, the newly established Enrico Fermi Chair in Physics. Among his many honors besides the Nobel Prize are the Albert Einstein Commemorative Award of Yeshiva University in 1957.

C. N. YANG—EARLY LIFE

Chen Ning Yang, the oldest of five children, was born in Hofei, China, on September 22, 1922. His father, Ke Chuan Yang, was professor of mathematics at Tsinghua University near Peiping and his son grew up in the nourishing atmosphere of a college community.

Early in life he was fascinated by the great mysteries and paradoxes of the cosmos—perhaps because of his upbringing in the mysterious environment of Taoist China, that great religion where common sense is stood on its head; where Occidentals describe a man as walking, the Chinese Taoists describe the event as "walking that is man-shaped." This focus on the event rather than the thing would ultimately find its way into modern physics with the advent of quantum mechanics, where waves are "really" particles and vice versa. This great element of almost religious mystery would soon lead him, along with T.-D. Lee, to the very summit of world physics.

After secondary school, Yang attended the National Southwest Associated University in Kunming, China, receiving his B.S. degree in 1942. Next, he attended Tsinghua University and earned his M.S. in 1944. In 1946, at the end of World War II, Yang went to the University of Chicago to study under the great U.S. physicist Edward Teller, though he was also greatly impressed by the pioneering work in nuclear physics by Enrico Fermi, also at the University of Chicago at the time.

He earned his Ph.D. in 1948 and served as an instructor at the University of Chicago until 1949, when he seized the chance to attend the Institute for Advanced Study in Princeton. It was at Princeton that he began his study of the weak interactions between the particles within the atomic nucleus.

Elementary particle physics divides particle interactions into several types, including strong and weak. Weak interactions are responsible for nuclear reactions. Meson decay, or the breaking up of mesons (particles that carry the "strong" forces within the nucleus) into their constituents, is one such example. It was this work that ultimately led him and Lee to their universe-shaking discovery of parity violations.

Moving to the State University of New York at Stony Brook in 1965, he continued his research on elementary particles as well as statistical mechanics. Most importantly, he began even more penetrating research into questions of symmetry in science—an interest dating back to his undergraduate days in China where he had written a thesis dealing with symmetry, titled "Group Theory and Molecular Spectra." His master's thesis was titled "Contributions to the Statistical Theory of Order–Disorder Transformations," and many of the ideas in it contributed to his Nobel Prize winning work later on questions of parity.

Many of the early papers written on parity showed that the parity of the pi meson, merely one type of "meson," is odd and that its angular momentum (the force generated by an object moving in a circle, which increases with its speed) was zero. While this did not prove parity was violated, it became important when Yang found himself enmeshed in the same problem that had captivated Lee. He began analyzing the decay of still another carrier of the "strong" forces in the nucleus, the so-called "κ meson," and it was that particular analysis that led to the apocalyptic discovery that in the weak interactions, parity is not conserved.

In 1957 he was awarded the Albert Einstein Commemorative Award and in 1958 he received an honorary doctor of science degree from Princeton University. He is a fellow of the American

Physical Society and the Academia Sinica, the prestigious Oriental analogue to the American Physical Society, as well as the National Academy of Sciences.

THE HISTORICAL IMPORTANCE OF SYMMETRY IN SCIENCE

The structure of cells and various biological processes exhibit symmetry. In mitotic division in cells, the chromosomes line up and are perfectly symmetrically placed prior to cell division. Einstein himself was guided by considerations of symmetry in his work on relativity theory. His assumption that space would appear the same for observers anywhere under his specified conditions demonstrates this phenomenon convincingly.

Still another example of symmetry appears in Einstein's later, "general" theory of relativity, where he proves the "principle of equivalence." According to the latter, an "accelerating" frame of reference is equivalent to an "inertial" reference frame, or a frame of reference where everything is moving at a constant velocity. In the latter, space and time must appear the same for all observers. In other words, gravity cannot be distinguished from inertia. (In simpler terms, a blind person falling from an airplane could not tell if he were falling, or merely floating in a weightless environment.)

Another way to illustrate the idea of symmetry or parity is to say that if you were to watch any physical event (say a rubber ball bouncing on the floor), there should be no way to tell, from merely watching the ball, whether you were looking at the ball directly, or watching it in a mirror. More picturesquely, one might say that in Alice's "looking-glass" world, the idea of parity tells us that the laws of physics should work the same way in or out of the looking glass. Thus, a ball dropped to the floor in the world of the looking glass should bounce in exactly the same way as it does in the "real" world. An apple would drop to the floor just as quickly inside the looking glass as in the "real" world. In physics, this is called the principle of "reflection invariance."

Yang and Lee proved that this was not so. Alice's looking-glass world really would behave differently from our own. That is why "looking-glass" milk is poisonous.

Before Yang and Lee's discovery, scientists believed that symmetry was critically related to several principles of *conservation*. A conservation principle, stated very generally, says merely that when something undergoes a change, certain things about it nevertheless do *not* change. Thus, for example, when mass is converted to energy, the amount of energy produced should equal the original amount of mass. (When dough is changed into cookies, the mass of the cookies should equal the mass of the original dough.) An apple in a left-handed universe would taste just as good as one in a right-handed universe; an electron in a left-handed universe would have to be the same size as one in a right-handed universe, and so forth.

Before the discovery of parity violations, physicists believed that *any* physical property would be "conserved" in this way. "Spin" was another such example. In any experiment, there would be no experimental way by which one could distinguish a particle spinning to the right from an identical particle spinning to the left.

The discovery of this conservation principle convinced generations of scientists and philosophers that nature would always respect symmetry. Indeed, scientists often guided their research based on the assumption that they would *always* find symmetry.

CHALLENGES TO SYMMETRY

In this era, this belief had been powerfully challenged. For example, Linus Pauling discovered in the 1930s that there were asymmetries in the structure of protein molecules.[2] In all probability, the most dramatic example came from his laboratory at Caltech. For example, he and many others believed that protein had a helical structure (much like the familiar "winding staircase" of the DNA molecule). But while other biologists believed that this helix would repeat itself in fully integral sequences, Pauling alone realized that the helix could have a nonintegral and therefore *asymmetrical* number of turns or repetitions of each amino acid along the helical chain. It could repeat itself two-thirds of the previous distance one time, 1.36 times the original distance the next, and so forth. (Sort of like the paradigmatic 1.6 children of the "average" American.)

Other disturbances in the idea of a completely orderly cosmos have appeared as well. Closely related to symmetry is the idea of certain knowledge. For centuries, scientists believed that it was always possible, at least in principle, to have absolutely certain knowledge of the workings of the universe, its laws and operations. The idea of certainty has been challenged by Heisenberg's "uncertainty principle" and quantum mechanics, both of which are now widely accepted. Many now believe that our knowledge of the cosmos must always be merely probable, a bothersome development both in physics and in theology. For this is regarded by many as a sign of disorder and imperfection in the universe, the sort of thing a perfectly Benevolent Grand Designer would or should not allow. Where, for example, classical physics said we could know that we did lock the car door, quantum mechanics says we can have only "very probable" knowledge that we did so (a theoretical encouragement to car thieves, I suppose, but not likely to be of much practical help).

More directly related to Yang and Lee's interests were the studies of particles conducted in 1953, where they began to probe in far greater depth the odd behavior they had chanced upon years earlier. In that year, the study of a type of subatomic particle called a meson hinted for the first time that right and left handedness might not be indistinguishable in nature. They and other scientists studied the behavior of two K mesons, called theta and tau. Though they appeared to have identical masses, identical life times, and were in fact in every other way identical, they nonetheless decayed in different ways. That is, tau decayed into three pi mesons, while theta decayed into two pi mesons. In other words, except for the way they decayed, they were in all respects identical twins! But this was a great paradox: How could two particles—identical and indistinguishable in every possible respect—nevertheless decay in two different ways? As Yang and Lee saw things, the only solution was to accept the idea that parity did not hold.

Thus, in the 1950s, both began to doubt the law of parity conservation. Instead, Lee and Yang believed that the so-called "weak" interactions described above do not follow parity. (Along with gravity, there are several other basic forces in nature: gravity is the feeblest one known.

Lee and Yang knew that the principle of conservation of parity had never actually been proven, although experiments had seemed to support it. Their suspicions now greatly aroused, they dug even deeper. They theorized that a subatomic particle called a "pi meson" might break apart or "decay" into two other particles. They suspected also that the particles resulting from the breaking apart of the pi meson might "prefer" to spin in one direction only.

Did they? Again, the classical parity law said that the resulting particles would have to spin in opposite directions, if only because the opposite spins would then exactly cancel one another. In fact, the exact reverse was seen in an experiment performed at Columbia University in 1957 with the famed Nevis cyclotron. The resulting particles were, in fact, spinning *only* left-handedly—not right—thereby violating parity. Because of this, they were awarded the Nobel Prize in 1957.

The clearest proof of a violation of conservation of parity was, however, seen in an experiment performed by C. S. Wu and her team, also at Columbia, in a study of beta decay of cobalt-60. In this work, Wu theorized that when cobalt nuclei were aligned in a magnetic field, they should give off as many electrons parallel to their spin as opposite to it—if parity was in fact conserved. This symmetry could be observed, again according to parity, merely by looking at the emission of electrons in a mirror, since parity demands that we see the same thing whether looking at the process itself or its mirror image. However, if the electron emissions in the direction of spin differed from the emissions in the opposite direction, they would not see the same thing in the mirror. The latter is exactly what Wu observed. Thus, we now know that neutrinos are always spinning left-, rather than right-handedly, and vice versa for antineutrinos—similar to what is observed in nature regarding the protein molecules.

To say physics was shaken by the discovery would be an understatement. To mention just one scare, in 1926 Fermi carried out his famous calculations on how atomic nuclei break apart by ejecting electrons, a process now called "beta decay." During the summer of 1937, a conference on Fermi's work on beta decay was held at Cornell University. The conference seemed to show that Fermi's theories were roughly correct. But, for a variety of reasons, after the news of Yang and Lee's work emerged, Fermi's ideas were suddenly in grave doubt, since they appeared to depend on the parity law. Eventually, parity was shown to be inessential to Fermi's work. Of course, the Italian physicist was greatly relieved.

BIOLOGICAL SYMMETRY AND ASYMMETRY

Further asymmetries are now known to exist. The molecules composing the chemical structure of milk, fortunately for Alice, occur only in left-handed forms, rather than the poisonous right-handed form. That is, they will steer light waves to the left, rather than to the right. (Sort of like entering a revolving door: if you enter on the left, the door steers you to the right and vice versa.) In nature, all proteins are built up of spirally arranged strings of units called amino acids. Virtually all of these, with the exception of the penicillin molecule, exist only in the left-handed form. (Scientists speculate that perhaps this is why penicillin is poisonous to bacteria.)

However, any right-handed protein would display exactly the same *chemical*, if not biological, behavior as its left-handed counterpart. Although right-handed forms almost never exist in nature, it is possible to make them synthetically, merely to verify the physical properties they are predicted to have.

A further oddity is the fact that while all naturally occurring protein molecules have only left-handed forms, when they are synthesized in the laboratory, both right-handed and left-handed forms will exist in approximately equal portions in the final product.

Why this is the case is not clear. One possibility is that at the dawn of evolution, the most primitive plant and animal species did contain both right- and left-handed proteins. But, over the course of evolution, it may have turned out that the left-handed form was hardier than the right-handed one. Thus, natural selection could have operated to gradually select against the right-handed forms, which therefore eventually all but died out.

In a word, nature prefers left over right, a consolation to left-handed people everywhere.

CHAPTER 16

The Emergence of Solid-State Physics

Solid-state physics, or as it is often called today, "condensed matter" physics, is the science of the properties of liquids and solids. Since it is less glamorous than, say, cosmology, the public has barely noticed it. Even physicists themselves have only slowly appreciated its importance. Yet this indifferent attitude is puzzling. In fact, the study of the solid state is replete with bizarre and extraordinary occurrences. Some metals, for example, at very low temperatures display a bewildering array of properties including superconductivity, the striking ability of some substances to lose all resistance to the passage of electrical current.

Another astonishing phenomenon is "superfluidity"—the uncanny ability of some substances to flow virtually resistance-free. (The study of this latter quasi-miracle is among the many accomplishments of the legendary P. W. Anderson of Princeton.) In recent years, fortunately, this neglect has at least started to ease. Benefiting from the quantum revolution, studies into defects in crystals, band theory (closely grouped energy levels in atoms), superfluidity, super- and semiconductivity have finally appeared.

EARLY WORK

It is true that some work in solid-state physics dates back to antiquity. The study of crystals, for example, is hardly new: Arab scientists or "faylasufs" studied and classified crystals centuries ago, believing at first that crystals were merely a rather rare type of solid.[1]

Today, with the exception of glass, we know that the great majority of solids are crystals. Besides common table salt, there is quartz, diamond, microcline, and many others. Each crystal has its own characteristic shape and structure. The architecture of quartz, for example, emerges from very large numbers of hexagonal crystals fabricated out of atoms of silicon and oxygen.

The well-known symmetry in crystals is due to the fact that the "unit cell," or the basic structure, repeats over and over again in a parallel sequence throughout the substance. This forms a "lattice"—a highly ordered repetition of the unit cell.

THE ADVENT OF X-RAY STUDIES

In 1912, a young physicist named Max von Laue of the University of Munich, along with his German physicist colleagues Walter Friedrich and C. Paul Knipping, devised a technique now called X-ray crystallography. The fundamental principles of X-ray crystallography depended on the fact that an orderly array of atoms will scatter X rays onto a photographic plate in a regular and predictable fashion. From this X-ray "footprint" on the plate, scientists could reason backwards to deduce what structure or arrangement of atoms the substance must have had to scatter the X rays as it did. In principle, the information derived could be vast and formidable—distances between atoms, sizes of angles between chemical bonds, and many other aspects of structure might be gleaned.

When the new technique appeared, information came pouring forth. Scientists were suddenly able to at least attempt useful explanations of the chemical and physical behavior of crystals. In fact, it could be argued that this technique alone transformed solid-state physics into a mature, respectable branch of physics.

THE STRUCTURE OF CRYSTALS

Scientists, realizing the importance of this work, carried it over into the realm of atomic physics, where it offered some of the earliest support for the famous Bohr model of the atom, later modified by German physicist Arnold Sommerfeld. Because of this support, interest in solid-state physics soared throughout Europe. Later, Dorothy Wrinch, Francis Crick, and others would apply X-ray crystallography to the structure of hemoglobin, sickle-cell anemia, and the structure of proteins.

ROSCOE DICKINSON'S WORK

Soon, the technique found its way to America through the efforts of Roscoe Dickinson as well as others at the California Institute of Technology. Dickinson had been Pauling's advisor from his first days in Pasadena and was one of the consummate geniuses in X-ray crystallography. Progress came gradually, as one substance after another yielded to Pauling's and Dickinson's relentless probing. Microcline, labradorite, and many other crystals revealed their structures over the next several years. Memories at Caltech were strong regarding Pauling's and Dickinson's early research in this field. As Professor Richard Hughes, professor emeritus of chemistry and a long-time colleague of Pauling's in the latter's protein studies, told me in his office at Caltech in 1984:

> Pauling did crystal structure from the very beginning as a graduate student. He would guess what the structure might be like, and then he would arrange it to fit into the other data . . . he could then calculate the intensities [of the X-ray picture] . . . he would get that

structure and then compare it with the observed ones. . . . That was his way of doing things.

Pauling himself discussed his work:

The research that I began was on the structure of crystals by X-ray diffraction. It was a rather new technique then. CIT [Caltech] was the place where the first American work in this field was done. That was in 1917. They [the early pioneers in diffraction studies] had continued their work there, especially Professor Dickinson, and he was the person that I worked with.[2]

The study of crystals perhaps reached its zenith with prominent British X-ray crystallographer W. L. Bragg's arrangement of an international conference of crystallographers in 1946 in London. Soon after that, interested crystallographers formed the International Union of Crystallography on August 3, 1948, at another conference held at Harvard University, resulting in the creation of new offices and new journals—principally *Acta Crystallographica*, the "Bible" of crystallographers.

METALS

As of 1935, scientists only superficially understood the structure and the chemical behavior of metals. Back in 1916, one of the distinguished early European solid-state physicists, H. A. Lorentz of the University of Leyden, had proposed the theory that metals were characterized by the ability of some electrons to "roam" freely through them, an idea further elaborated upon by Sommerfeld, Fermi, and several others. Still, the theory could not fully explain some of the metallic and physical properties of mixtures of metals, or alloys and boranes (combinations of boron and hydrogen).

Though many of his ideas are no longer accepted, Pauling ventured into this area of research when he suggested, based on his observations of the properties of metals and what might be required to explain them, that there might be an extra "metallic" orbital. By theorizing the existence of these extra orbitals, he could account for the great conductivity of metals. Such extra electrons are able to roam freely through the metal, thereby allowing electrical conductivity. Pauling comments on this in a May 16, 1949 letter to his old mentor Arnold Sommerfeld, saying ". . . I have published . . . a theory of metals . . . based on the idea of the resonating valence bonds, is rather a complex one . . . and I have succeeded in explaining away many of the properties of a metallic system on this basis."

VAN VLECK'S CONTRIBUTIONS

No discussion of solid-state physics, quantum chemistry, crystal structure, and especially magnetism, one of the most familiar and dramatic properties of solids, could be complete without some mention of John H. Van Vleck. He was born in 1899 in Middleton, Connecticut, the son of distinguished mathematician Edward Burr Van Vleck. He received his bachelor's degree from the University of Wisconsin in 1920 and later studied physics under the fabled P. W. Bridgman, professor of physics and philosophy at Harvard. Eventually he became professor of physics at the University of Wisconsin. During World War II, Van Vleck worked at the MIT Radiation Laboratory and Arthur Compton later invited him to work on the National Academy of Science's committee to study nuclear fission. After the war's end, Harvard appointed him chairman of the physics department. Later, in the early 1950s he became the first dean of engineering and applied physics at Harvard. In 1951, he became Hollis Professor of Mathematics and Natural Philosophy at Harvard.

Van Vleck did much important work in "quantum chemistry"—showing how the principles of quantum mechanics could clarify problems in chemical bonding. That is just one area where Van Vleck helped "reconcile" two apparently disparate points of view. In particular, he did much to "reconcile" the "molecular orbital" and "valence bond" approaches to understanding chemical bonding. In the former, the electrons are thought of as having a path or "orbital" influenced by two or more nuclei of atoms bound together; but in the valence bond theory, the orbital paths of the electrons are thought of as being located around single atoms.

He also developed the *crystal field theory*, showing that a close study of the behavior of electrons in crystals could yield a greater knowledge of their energy. In this way, Van Vleck explained such ordinary properties of crystals as electrical conductivity and their optical qualities, such as the ability to reflect and refract light.[3]

WORK IN MAGNETISM

Van Vleck's important work in magnetism began in 1926, when he showed how the then embryonic science of wave mechanics could explain how various substances were affected by magnetic fields. In 1932, that wonderful year of the discovery of the positron, the neutron, and heavy hydrogen, Van Vleck published *The Theory of Electric and Magnetic Susceptibilities*, applying quantum mechanics to ferromagnetism (ordinary magnetism, or the capability of substances like iron to become magnetized by any weak magnetic field). With this work, he joined such giants as John Slater in studying these problems. He is perhaps best known for his application of matrix mechanics (a form of quantum mechanics) to the study of the magnetic properties of a wide assortment of substances.[4] Deservedly, therefore, he is known as the "father of magnetism." Scientists soon applied his work to practical problems in the development of lasers as well as to magnetic resonance imaging. The latter is a method of producing images of the body using a combination of a magnetic field and radio waves. Physicians and hospitals make wide use of the latter in diagnosis.

Van Vleck shared the 1977 Nobel Prize with physicist Nevill Mott of Cambridge University and the great Philip W. Anderson of Princeton University for ". . . fundamental theoretical investigations of the electronic structure of magnetic and disordered systems."

After this, his health began to deteriorate and he gradually scaled down his scientific work. He died of a heart failure in 1980.

PHILOSOPHER AND SCIENTIST P. W. BRIDGMAN

If Van Vleck was among the most practical of America's scientists, Bridgman was, save possibly for Einstein, the most philosophical of physicists. Indeed, he accomplished memorable work in both areas. Best known for "operationalism," he followed in the pragmatist tradition of Dewey, Peirce, and James in the past and Harvard's Israel Scheffler and the University of Virginia's Richard Rorty today. Thus, he rejected as "meaningless" any concept that could not be defined in terms of observable laboratory procedures. He worked this concept out in *The Logic of Modern Physics*, published in 1927. Much dissatisfaction with the then-current physics on his part prompted this work. For one thing, he disliked the highly rarefied metaphysical, and therefore untestable ideas found in the field of electrodynamics, or the science that studies the mechanical forces generated by circuits carrying electrical currents. He influenced many, although his influence was greater among experimental rather than theoretical physicists. More importantly, he influenced a generation of philosophers working in the philosophy of science, although it is pretty widely agreed that his "operationalism" is a bit too restrictive. As it turns out, there are many useful concepts in physics that cannot be easily operationalized (such as the idea of a quark, or "antimatter").

In 1929, Washington University physicists E. U. Condon and P. M. Morse wrote of him, "In thinking about all physics, . . . the operational point of view stressed by Bridgman in his *The Logic of Modern Physics* is a most important aid."[5]

Bridgman was born in Cambridge, Massachusetts. He received his B.A. from Harvard in 1904 and his Ph.D. in 1908, for a dissertation on the effect of mercury's resistance to electrical current. A principal source of philosophical influence came from his Harvard association with A. N. Whitehead, who was renowned for his work with Russell, a distinguished British philosopher known primarily for his development of logical atomism—the doctrine that stresses the critical importance of verification in understanding language and for reducing mathematics to a few principles of logic (a sort of "grand unified field theory" of mathematics).

In personality, Bridgman most resembled the great A. A. Michelson. Both were great experimentalists and neither had much loyalty to school or graduate students. As Bridgman once said to a dean, "I am not interested in your college, I want to do research."

SOLIDS UNDER PRESSURE

Doubtless his finest contributions to the study of matter in the solid state were his investigations into the behavior of solids under extreme pressure—specifically inordinately high pressures. Among his most enchanting results was the discovery of "hot ice"—a form of ice that does not melt even up to 180°F, as long as it is under a pressure of nearly 300,000 pounds per square inch. In 1946, the Nobel Committee awarded him the Nobel Prize in physics "for the invention of an apparatus to provide extremely high pressures and for the discoveries he made therewith in the field of high-pressure physics. . . ."

After a diagnosis of bone cancer, Bridgman committed suicide on August 20, 1961. A philosopher to the end, the suicide note read "It isn't decent for society to make a man do this . . . himself. . . . Probably this is the last day that I will be able to do this!"[6] In a quarter of a century, the "right-to-die" issue would become a fiercely debated conundrum in medical ethics.

JOHN SLATER'S SCHOOL OF SOLID-STATE PHYSICS

Some physicists seem to feel that John Slater may well be most notoriously remembered for the *way* he did physics, rather than for his actual contributions (although these too are formidable). Certainly he was among the most conservative theoretical physicists America ever produced—afraid to suggest a theory or explanation until he had done every imaginable calculation. Nobel laureate P. W. Anderson once described Slater well by referring to his "crank-turning, grind out all the electrons. . ." method of doing physics.[7] Professor Sten Samson, in an interview at Caltech, seemed to agree with this evaluation of Slater, saying, "while that way of doing physics can produce good results, there is also a danger of losing the forest for the trees."

Still, Slater had supporters. In a letter to Peter Adams, editor of *The Physical Review* in 1984, a disappointed author takes a swipe at Anderson and offers implicit support for Slater's approach to physics:

> Anderson is a great physicist, but he seems to have a deeply rooted prejudice against the style of physics that could be generally labelled Slater-like. The article written by Anderson for *La Recherche* . . . is a diatribe against "automatic machinery which will produce the answers without thinking," and the "disease of physics these days, and apparently our kind of physics, is the idea that results can be had with some kind of automatic procedure, which involves only drudgery" . . . .[8]

Was Slater really as thorough as his reputation seems to indicate? In fact, a close study of Slater's correspondence does show many references to his painstaking calculations. In a letter of December 18, 1959, for example, he says:

> It's too bad that the number of configurations goes up so astronomically as the number of atoms increases, but with the computing facilities we have now, I think we should go as far as we can, and get some quantitative information as to the nature of electronic energy levels . . . as a function of internuclear distances more complicated than the hydrogen molecule.

Earlier, in a letter of April 11, 1931, he says, "A student here is working out exchange integrals, and others, so that perhaps by June there may be something more quantitative to report."

When all was said and done, Slater got results. By the 1950s, better computers were available and Slater's group was starting to make brisk progress with their quantum mechanical calculations. A revolution occurred when the IBM 704 appeared near the end of the 1950s. With this, Slater progressed rapidly his now-famous work on *energy bands* (closely grouped energy levels in atoms). This is now one of the premier theoretical approaches to the study of the properties of matter in the solid state. Slater, building on the work of Wigner and Seitz, was able to compute the excited band widths of sodium electrons, while many of his students continued this work with other metals.[9] The importance of this work lay in the fact that the nature of bands determines such physical properties of a substance as electrical conductivity. In a metal, for example, the bands have plenty of unoccupied space, so that electrons can flow freely through them. That is why metals are good conductors of electricity. (Electrons in metal bands are like adjacent auditoriums with relatively few people in each— electrons can flow freely through metals just as people in such circumstances can wander freely between rooms.)

SLATER: EARLY LIFE

Slater was born in 1900 in Oak Park, Illinois. He received his bachelor's from the University of Rochester and completed his doctoral work under P. W. Bridgman at Harvard in 1920. He spent a postdoctoral year in Europe under a Sheldon Traveling Fellowship presented to him by Harvard. He became the department chair at MIT in 1930, remaining there until his retirement in 1966.

Slater's paper, "The Theory of Complex Spectra," published in 1929, offered one of the most profound concepts so far given in quantum physics, the so-called "Slater determinant," a technique that powerfully influenced the work of numerous other scientists, including Sten Samson and E. U. Condon.[10] A rigorous mathematical idea, it greatly simplified the mathematical aspects of the study of the structure of the atom.

SLATER'S WORK ON MAGNETISM

Slater also did considerable work on magnetism, building on previous work by Heisenberg, Pauli, and others. Pauli's work influenced many; he deepened our understanding of magnetism by applying Fermi statistics (dealing with the motions of electrons within metals) to magnetism, and showing the relationship of magnetism to temperature.

Slater's work here is, however, difficult to evaluate. As Nobel laureate P. W. Anderson wrote me on September 21, 1988,

> . . . Slater's views on magnetism are at best controversial [because many physicists do not believe that Slater's views on the electronic structure of metals fully explain such metallic properties as ferromagnetism]. The modern theory of metals, culminating in the work of Migdal, Landau, Hein (for magnetic metals, Herring), Cohen, and many others, . . . is a spectacularly accurate and formally rigorous intellectual structure. . . .

SLATER'S WORK ON VALENCE

His contribution to chemical bonding is much clearer. In fact, he contributed a great deal to the development of *valence bond theory*, one of the two major approaches to understanding chemical bonding. Among others, German physical chemist Walter Heitler influenced him. In a letter of April 11, 1931, he says, "Heitler's paper also has just come, and I must study it. The thing I have had on my program is to work out my valence ideas . . . and I have been holding off because I knew Heitler's paper was on the way. . . .[11]

He died in 1976, but remained active in physics throughout his life. Despite the controversy noted above, his research into the nature of metals, band studies, the nature of magnetism, bonding, and so forth at least fueled a strong and continuing interest in solid-state research—studies that now promise a virtual technological utopia. It is to one of these exciting new frontiers in solid-state physics that we now turn.

SUPERCONDUCTIVITY

The phenomenon of superconductivity was first discovered by Heike Kamerlingh Onnes in 1911 at Leiden, while he was studying the conductivity of mercury at low temperatures. The theory of the time foresaw that as the ordinary vibrations of metallic atoms decreased with ever lower temperatures, resistance to electrical current would shrink as well. But, at 4.12 K, or about 455° below 0°F, near absolute zero, a remarkable phenomenon occurred—the resistance of the mercury to an electrical current vanished completely. Once the current starts, it will go on virtually forever. As such, it becomes a fantastic source of power and it is little wonder that so many are interested in it today. There are many practical applications. Besides their use in building particle accelerators, superconductors are used to construct the powerful magnets found in the magnetic resonance imaging equipment used in medicine. Also, future plans for elevated trains call for superconducting substances.

Soon, scientists realized that other metals were capable of this "superconductivity." By dropping the temperature to 3.73 K, tin became superconductive; lead at 7.22 K, uranium at 8 K, aluminum at 1.2 K, and, lowest of all, hafnium at only 0.35 K. Whether other metals can be made superconductive at any temperature is a project for the future.

ORDINARY CONDUCTIVITY

In order to understand superconductivity, it is necessary to understand the structure of metals generally. Metals are characterized by the fact that some electrons, not tied to any one atom, can "roam" freely through the metal, thereby allowing the conduction of electricity. Consider two atoms next to each other in a metal: Because the two atoms are "jammed" so close together, the nucleus of one atom will actually be inside the outer electron shells of its neighboring atom in the metal. As a result, the electrons in these outer shells will be attracted by the nucleus of the "intruder" neighboring atom. Thus, these electrons are "pulled" and can spread throughout the metal, thereby conducting electricity. (In substances like table salt, by contrast, the electrons, even the outer ones, are tied too rigidly to their atoms to roam like this. Thus, salt, or any substance like it in this respect, cannot conduct electricity. In fact, such substances are *insulators*—the opposite of a conductor.)

Science reached an important plateau in understanding the phenomenon in 1957. In that year John Bardeen (who also made important contributions to understanding *semi*conductors, or substances, such as germanium, intermediate in conductivity between conductors and insulators) and his colleagues J. R. Schrieffer and L. N. Cooper published a theory known as the "BCS" theory. With the latter they managed to explain most of the experimental results connected with superconductivity, as well as predict new ones.[12]

An important feature of superconductivity is electron "pairing." In the superconducting state, the metallic structure is unusually highly organized and therefore extremely stable. In this state, the metal tends to lose any outer electrons, since they are not as tightly bound to the nucleus as atoms in the inner shells. The outer electrons can then wander throughout the metal. But there are certain restrictions, because of the principles of quantum mechanics. For example, electrons cannot occupy just any location. They can only make discontinuous "jumps" from one level to another and cannot occupy intermediate points, just as a child walking downstairs can be on one step or the next below it, but has no place to stand in between.

The most stable arrangement of electrons is one where all of the lowest or inner energy states are filled, and the highest or outer ones are all empty. If a current is then induced, an electron pair forms. In other words, as a current—an electron—travels through the lattice, it attracts the positive charges in the lattice toward it. This creates an excess positive charge which, in turn, attracts still another electron. In effect, this second electron is indirectly "paired" with the first.

In superconductors, these pairs do not break up. The reason is that the temperature is so low that the electrons forming the pairs are almost vibrationless. And as long as nothing is done to break them up (i.e., raising the temperature of the metal over the superconducting point so that the electrons become so agitated that the pair breaks apart), the current will go on through all eternity. Another by-product of superconductivity is diamagnetism—the ability of superconductors to *repel* magnetic lines of force. German physicist Meissner pioneered this research in the early 1930s.[13]

By the 1950s and 1960s, P. W. Anderson of Princeton University and the Bell Telephone Laboratories had become interested in superconductivity. Although Bell Labs was aware of the possible practical applications of superconductivity in medicine, particle physics, and electronics generally as discussed earlier, Anderson had no practical applications in mind; the new interest in superconductivity grew out of his previous work. Earlier he had successfully explained how individual atoms of a magnetic substance such as iron could behave as tiny magnets—despite being embedded in a *non*magnetic material. (Iron atoms embedded in aluminum would be an example.) Anderson improved and amplified Bardeen and Cooper's work, showing that superconductivity was not an isolated phenomenon in substances, but rather was closely related to other properties of the substance in question. Perhaps most remarkable was Anderson's work in "dirty superconductors"—the study of the effects of various kinds of impurities on superconductivity.

Remarkable progress came in the 1980s when Anderson published a ground-breaking theory explaining how some substances could become superconductive at much higher temperatures than previously thought possible—perhaps even at room temperature.

In 1986, K. A. Muller and J. G. Bednorz found an oxide compound that would be superconducting at 30 degrees above absolute zero. Since that was the highest temperature ever found at which a material became superconducting, it offered great encouragement to seek out materials that would be superconducting at even higher temperatures. In 1987, scientists created the first "warm" superconductor—a material that becomes superconducting at only −321°F, the temperature of liquid nitrogen and the highest known temperature for superconductivity.

Then in 1988, researchers at the University of Arkansas found a thallium-based substance that became superconducting at −234°F. By this time, many scientists were also developing practical applications for superconductivity. Perhaps the most talked about application is in the realm of particle physics. The High-Energy Science Advisory Panel in 1983 suggested building the SSC, or "superconducting supercollider"—an H. G. Wells type of device over 50 miles in circumference.

However, given all of the problems with the Hubble Space Telescope, no one wanted to see these problems on the almost $10 billion SSC. So it was that SSC physicists, under the direction of Roy Schwitters, thoroughly examined the lesson from Hubble. If Congress does fund the SSC, there will be more than 2000 people employed on the project. As of this writing, the SSC is scheduled for completion in 1999. It will use 10,676 magnetic superconductors. Once in the collider tube, the researchers will then accelerate protons to very nearly the speed of light. That done, the team will initiate collisions with other elementary particles under conditions similar to those shortly after the big bang.

In 1988, Francis Moon and Rishi Raj managed to design a ball bearing that was almost entirely free of friction using a superconductor. But there are problems. Very often the materials that turn out to be superconducting are extremely fragile. Also, because of the nature of some ceramic-based superconductors, they cannot carry a very large current, though these problems have to some extent been overcome by designing "superconducting films." The great advantage of these is that their molecular makeup is different enough from the original superconducting material that they can carry somewhat larger currents.

SEMICONDUCTIVITY

Still another "glamorous" field in solid-state physics involves the study of *semiconductors*—substances whose ability to conduct electricity is between that of a conductor and an insulator. Their great application has, of course, been in the field of electronics, where semiconductor circuit "chips" have enabled manufacturers to greatly reduce the size of virtually any kind of electronic device, from VCRs to computers. Bell Labs did some of the earliest really important work in this field in the mid-1930s. In 1928, physicist Mervin Kelley had already spoken at MIT on solid-state telephone research. By 1935, Kelley was actively interested in working on solid-state amplifiers (a device for reproducing an electrical signal at greater intensity) and had already developed the rudiments of a theory of semiconductors, later to be perfected by William Shockley. Kelley organized a team consisting of physicists, metallurgists, chemists, and spectroscopists to study them.

Another critical stage in semiconductor research came after World War II. Some of the centers for this work were Bell Labs and John Slater's group at MIT. Because of the experience of physicists in engineering during the war, many returned to the Academy highly skilled in such practical techniques as radiofrequency spectroscopy, cyclotron resonance, paramagnetic resonance, etc.—all of which were critical tools for probing the internal structure of solids. Also, the development of computers in the postwar period aided researchers tremendously. With the emergence of quantum mechanics in the mid-1920s as a powerful investigative tool, computers were a necessity because of the enormously complicated calculations that were now required.[14]

Since germanium and silicon crystals were widely used in radar, scientists had already begun to understand these substances better. Research on these materials continued after the war, by Karl Lark-Horovitz at Purdue University, by Bell Labs (with William Shockley, John Bardeen, and many others), and by Slater's MIT team. Also, some of the students of Eugene Wigner (the German émigré from the University of Berlin) and Joseph Callaway at Bell Labs were able, by studying the energy bands of silicon and germanium, to accumulate much new information about the properties of these substances—most importantly on their semiconducting properties. Soon, and because of all of this work, scientists began to appreciate how useful substances like germanium and silicon could be as semiconductors.

SHOCKLEY ENTERS THE PICTURE

In 1936, Kelley hired William Shockley, surely one of the greatest solid-state physicists of modern times.[15] Shockley had also studied under John Slater. In fact, he, along with men like Eugene Wigner, John Bardeen, and Frederick Seitz, were among the first to be routinely described as "solid-state physicists." Shockley's first work at Bell Labs involved trying to improve vacuum tubes. Soon, however, he joined the research group of Harvey Fletcher, already distinguished for its work on acoustics.

Before long, Shockley, working with physicists John Bardeen and Walter Brattain of Bell Labs, ferreted out considerable information on the properties of semiconductors. To a great extent, this was due to the application of quantum mechanics to problems in the solid state. It was Shockley and his group, however, who actually produced experimental confirmation of many of the predictions of quantum theory. Shockley found, for example, that the efficiency of semiconductors varied considerably with temperature and other factors, such as the amount of impurities in them.

By 1947, Shockley and Brattain were able to put on a stunning demonstration of the first real electronic transistor, made out of semiconducting materials which replaced the old large and unwieldy triode tube seen in old radios and other old electronic equipment.

LIQUID HELIUM AND SEMICONDUCTIVITY

Still another critical advance in the study of solid-state crystals and their properties occurred in the 1950s, with the development at MIT of the Collins liquefier for liquid helium. The importance of such a device lay in the fact that the low temperatures reachable with this apparatus diminished the electrical resistance of substances considerably.

THE NATURE OF A SEMICONDUCTOR

In a semiconductor, such as germanium, all of the electron-energy levels are filled to the level of a band called the "valence band." Immediately above this, all energy levels are empty. The very top level of the filled energy gaps is called the "Fermi energy." On top of this, in turn, is a totally empty band called the "conduction band." The important "Fermi energy" lies inside this gap.

Now, with an ordinary metal conductor, such as copper, the Fermi energy lies not inside the gap as with a semiconductor, but inside one of the energy bands. With a semiconductor, charge is carried in two ways—by electrons or electron "holes."

The fundamental idea, for which Shockley must get considerable credit, is that an impurity in a substance will cause that substance to have more electrons than are required to bind the atoms of the substance together. These "excess" electrons are thus free to move as they please and in doing so, conduct electricity. A "hole" is essentially the converse. It is a region where electrons are *missing*. Since a negatively charged electron is missing, the "holes" carry positive charges and they too can in a sense "move." They "move" in that when one of these holes is filled by a forward-moving electron, it ceases to exist as a hole and a new "hole" appears behind it. Indeed, one of Shockley's major contributions was to further our understanding of how a "hole" creates electrical conductivity.

THE FIRST WORKABLE SEMICONDUCTOR

In 1947, the first really useful semiconductor appeared, consisting of germanium impurities that could generate the excess electrons and the holes which constituted and created the moving electrical charge. In 1950, Shockley improved on this idea by restructuring the semiconductor somewhat to produce what he called a "junction transistor"—a device that was easier to manufacture and far more durable. It would hardly be an exaggeration to say that his work made the modern computer era possible— including various kinds of medical technology, communications equipment of all sorts, VCRs, and countless other technologies. For this work, Shockley, Brattain, and Bardeen were awarded the Nobel Prize in 1956.

SHOCKLEY ENTERS THE WORLD OF THE INFAMOUS

After leaving Bell Labs in 1955, Shockley opened his own company, the Shockley Semiconductor Laboratory in Palo Alto, California, for the purpose of manufacturing and doing further transistor work. Missing the excitement of a college campus, however, he accepted a post at Stanford in 1963, and finally retired in 1975. After that his reputation began to slip, for he got himself tangled up in extremely controversial views concerning population control.

By the late 1970s, Shockley had retired from active social propagandizing though he always kept up with new scientific developments. He died in 1989.

For the future, work in developing semiconductors may take a drastic new course. Instead of the silicon-based transistors currently used in computers and other electronic equipment, comparatively inexpensive synthetic *diamond* could become an integral part of the new electronics. One reason for this is that circuits of the future may be light- rather than electron-based, since diamonds can transmit signals via light beams faster and with less interference than moving electrons can carry signals. Such "optical" computers are already being developed at Bell Labs but, as of this writing, are not commercially available.

Today, the cutting edge of solid-state physics has shifted a bit from transistor work to studies in superconductivity. IBM physicists K. Alex Muller and Georg Bednorz have, since 1987, reported success at developing superconducting materials at temperatures of almost 90 K. Since this is far above the boiling point of comparatively inexpensive liquid nitrogen, the latter can be profitably used as a refrigerant. Scientists at many laboratories are experimenting with other possible superconducting materials, using such substances as yttrium, copper, oxygen, and barium.

Once solid-state physicists solve the above-mentioned difficulties in making working superconductors, not only theoretical physics, but also our daily lives will be much easier. It is not impossible that, since superconductors repel magnetic forces, we will one day ride safely on superconducting, magnetically elevated trains as well as other types of vehicles.

CHAPTER 17

Murray Gell-Mann and More Particles: Forces within the Nucleus

MORE ANTIPARTICLES AND MORE PECULIAR BEHAVIOR

After Anderson's great discovery of the positron, or positive electron, by no means did the search for ever more fundamental particles stop. Instead, things got ever more complex. The list of particles continued to swell and some of them soon began to exhibit very peculiar properties which often contradicted known theory.

THE ANTIPROTON

It was not until over 20 years after the discovery of the positron, however, that American physicist Emilio Segrè and his team (Ypsilantis, Wiegand, and Chamberlin) discovered the *antiproton* during experiments with the Bevatron accelerator in Berkeley, a particle accelerator that could raise particles to an energy of several billion electron volts—far greater energies than had been produced by any previous cyclotron. Soon, this discovery led to the general conception of *antimatter*—entire atoms composed of nothing but antiparticles, antiprotons and antineutrons, all surrounded by shells of positrons. In a sense, these were the "mirror images" of ordinary atomic nuclei. (Doubtless this "parallel antiuniverse" was a boon to sci-fi writers.)

WHAT IS HOLDING THE NUCLEUS TOGETHER?

But while new particles were appearing, others were vanishing. At this point, what had once been a well-accepted theory of the nucleus had to be abandoned. Scientists had previously believed that electrons could exist inside the nucleus itself. Because of their negative charge and consequent affinity for protons, they could help serve to bind the nucleus together.

But, for various reasons, it turned out that such electrons needed to have such high energy levels that it would be impossible to hold them inside an ordinary sized nucleus under any normal conditions. In fact, the nucleus would have to be about 10^{-13} meters across, which it isn't. So the electron could not be the "glue" holding the nucleus together. In that case, what was holding the nucleus together? Whatever it was, it would have to be immensely powerful in order to overcome the repulsion between particles of like charge within the nucleus.

At this point, the famous quantum mechanical "uncertainty principle" of Werner Heisenberg comes into play. Stated in its usual form, the principle says that it is impossible, in principle, to measure both the position and momentum of a subatomic particle with absolute accuracy, though the uncertainty also exists with respect to other variables, such as energy and time. That is, by measuring the energy of a system exactly, one makes it impossible in principle to know how long the particle will last, and vice versa. (Why this is so is not a question within the scope of physics: the universe is simply built that way.)

THE STRONG FORCES

Thus, if we first measure *exactly* how long some process or event lasts ("event" being used in the ordinary sense to denote any process taking some amount of time), such as how long a particle lives, it is impossible to say exactly what energy the particle had during that time. The "uncertainty" factor would be infinite, and vice versa.

But it is this very same uncertainty that solves the problem of the strong nuclear forces. Once again, as has happened so many times since 1925, quantum mechanics came to the rescue of a seemingly insolvable problem in physics.

THE FIRST GLIMPSE OF THE STRONG FORCES

As is so often the case, strong hints of a breakthrough come before the breakthrough itself. In 1936, for example, a glimpse of the future came with the work of Heydenberg, Hafstad, and Tuve at the Brookhaven Laboratory. They conducted some of the first thorough investigations of proton scattering, an experimental technique where protons are systematically collided with one another. In so doing, the first preliminary hints of very strong, short-range, and charge-independent forces between protons appeared. The experimental work was complemented by theoretical work done at Princeton by physicists Condon, Gregory Breit, and Richard Present.

Now again, we know from Einstein's famous equation $E = mc^2$ that any amount of energy can be converted into an equivalent amount of mass and vice versa. Thus, the uncertainty relationship between time and energy just discussed can be extended; it can easily be stated in terms of an uncertainty relationship between time and mass. So, if we have a particle with a certain mass, and we take some nonzero amount of time to measure that mass, there is going to be an uncertainty in the measurement of the mass. As before, the relationship varies inversely. That is, as the uncertainty in the time needed to take the measurement is *decreased*, the uncertainty as to the amount of mass there *increases*.

The logic emerges—as we diminish the uncertainty in the time, the uncertainty in the measurement of the mass of the particle is so great that we cannot even be sure whether there is a *single* particle there or two! So, one neutron could become two (a "virtual" particle is created from the original) and then one again, or one proton could become two protons and then one again before we have completed the observation. *Thus, we could never detect the second "virtual" particle.*

But during the brief lifetime of one "virtual" particle, it could be exchanged for the virtual particle born of another real particle. For example, two protons could be "glued" together by exchanging each other's "virtual" particles. (It is known from work on quantum electrodynamics[1] that subatomic forces generally act via an exchange of particles.) Like children in a maternity ward, "virtual" particles can be switched at birth. Of course, this is all highly theoretical and cannot be directly "observed" in the sense in which a person robbing a bank can be observed, but it is the best explanation at hand of the forces holding the atomic nucleus together. In any case, the idea of "direct verification" really tends to become rather meaningless in the very abstract realms of theoretical physics. In a very real sense, theoretical physics, philosophy, and theology are coming much closer together as the 21st century approaches.

OTHER PARTICIPANTS IN THE CREATION OF STRONG FORCES

But that is not the end of the story. Neutrons also play a role in the "strong" forces holding the nucleus together. First, since they have no charge, at least they do not add to the electrostatic forces (forces resulting from electricity, such as is found in ordinary magnets) tending to break up the nucleus. But they do more than this; because they intervene between protons, they weaken the repulsive forces by pushing the protons further apart (except, of course, in the case of the *hydrogen* nucleus, which has only a single particle, the proton).

YUKAWA'S CONTRIBUTIONS

It was at this point that Hideki Yukawa's work comes into the picture. Working at Osaka University in 1934, Yukawa suggested that an exchange of a virtual particle occurs between two protons; in line with the above, this would result in an attraction between the protons.[2]

In Yukawa's seminal paper, he also showed that, under ordinary conditions (a nucleus in the core of a massive star would not be under "ordinary" conditions because of the enormous temperatures), this nuclear force could be powerful enough to overcome the electrostatic repulsion between the two positively charged protons. Yukawa further suggested that the range of the forces depended on the mass of the "virtual" particles. The more massive the particle, the shorter would be the range of its attractive force. Thus, the "strong" force, carried by very heavy particles, produces only a very short-range force. Beyond a specified distance, the attractive forces quickly drop to zero. On the other hand, the electromagnetic force—carried by photons, the particles composing visible and invisible light which have no mass at all—has an infinite range, as simple arithmetic shows.

All of the evidence pointed very strongly to the existence of yet another undiscovered particle, one that would, unlike the electron, fit the requirements for a nuclear force. Therefore, Yukawa suggested a particle that has come to be known as the "meson" (from the Greek "mesos," for "middle"). This particle would be far heavier than an electron, but much lighter than a proton. His theoretical speculations convinced him that such a particle would be strong enough to hold the nucleus together, but still allow a nucleus to have a size comparable to that of known nuclei.

As it turned out, the idea of virtual particle exchange had very wide applicability, though some of its applications have not yet been experimentally confirmed. Other applications have been well confirmed. It is known that the process of virtual particle exchange holds quarks, or the particles making up protons and neutrons, together. Also, it holds the protons and neutrons themselves together within the nucleus. However, the gravitational force, for example, has been theorized to result from the exchange of still another particle, the "graviton," though this one has not been found in nature. However, there is little doubt that mesons exist.

ANDERSON REENTERS THE PICTURE AND THE PARTICLES PROLIFERATE SOME MORE

Soon, Yukawa began to believe that perhaps his theory was vindicated. Again, the experimental genius of C. D. Anderson and his colleagues found a particle in cosmic rays with a mass of about 200 times that of the electron.[3] In 1938, Anderson, along with his colleague Neddermeyer, dubbed his discovery a mu meson. Naturally, there was widespread speculation that this might be Yukawa's meson. But unfortunately for Yukawa, a number of Italian physicists at the University of Rome refuted this theory. The team of Pancini, Piccioni, and Conversi showed, in 1945, that the forces exerted by the particles found in the cosmic rays simply were not strong enough to hold any nucleus together.

Soon, however, American physicists Hans Bethe and Marshak as well as other theorists in Italy and Japan unearthed a marginally heavier meson, which they promptly dubbed the "pi" meson, or "pion" as it is now commonly called, and it is considered a type of meson. The pion, it turned out, did generate enough force to hold a nucleus together. (The particle of the cosmic rays was then named the mu meson or "muon.") There was, finally, a link between the two, in that Yukawa's nuclear meson or "pion" decayed into the cosmic-ray meson.

THE HADRONS: A COMPLEX FILING PROBLEM

But the complexities didn't end here. The pi meson itself proved more complex than anyone had appreciated. As it turned out, there were three types of pi mesons. Mu mesons also had to be further categorized, as it turned out that there was both a muon and an antimuon. Researchers further realized that both muons and pions have extremely small lifespans and are rarely found in nature. Whatever muons can be found are usually found at sea level—the result of cosmic ray[4] bombardment of the atmosphere.

By 1947, various laboratories around the world had found still more new particles in cosmic rays. Classification was difficult however. Although they resembled some of the previously known particles insofar as they seemed to be a factor in the strong intranuclear forces, they existed only in pairs.

A CAPSULE SUMMARY

Nonetheless, physicists eventually managed to classify all of the known particles. Those not involved with the strong nuclear forces are neutrinos, electrons, and muons, which are collectively known as "leptons." (The only other forces importantly associated with these particles are electromagnetic forces and gravitation.) All other particles, neutrons, pions, baryons, mesons, protons, and many, many others came to be known as "hadrons"—the particles involved in the strong forces in the nucleus and which have a definite size and mass.

PARTICLES, PARTICLES, AND MORE PARTICLES

By the 1950s, with increasing interest in particle physics and with the advent of new and more sophisticated particle detectors such as the bubble chamber (a more sophisticated type of cloud chamber, in which the tracks of particles become visible) as well as the enormous cyclotrons such as the Bevatron at Berkeley, still odder and newer varieties of particles appeared. These too seemed to be important factors in the strong forces in the nucleus.

QUARKS

By the early 1960s, where there were once only protons, neutrons, and electrons, there were now hundreds of particles with an assortment of odd qualities. In the early 1960s, colleagues of Richard Feynman at Caltech, Murray Gell-Mann and George Zweig, suggested that all of the above "hadrons" might really be ultimately composed of an even more fundamental type of particle called a *quark* (Gell-Mann's term, taken from Joyce's *Finnegan's Wake*—"Three Quarks for Muster Mark"). The difference between quarks and hadrons, so far as is known at present, is that quarks cannot exist by themselves.

At first, Gell-Mann and his co-workers expected to find no more than three types of quarks. Their predicted charges were 2/3 and −1/3, which Gell-Mann in a fit of archetypal scientific mischievousness named "up," "down," and "strange," respectively, the latter denoting the quality or property related to the speed at which they decay. It was also Gell-Mann who suggested that the new quality of "strangeness" should be regarded as a new quantum number. Soon he could add a new *conservation* law to the textbooks, along with the laws of conservation of energy and momentum. It was soon shown that "strangeness" was conserved in all particle interactions except the weak ones, meaning merely that no matter how these particles interact with one another, the new property of "strangeness" would not be lost after the interactions were over. (Except, again, in the weak interactions.)

But Gell-Mann's guess of three quarks was too low; soon, experiments at the Brookhaven National Laboratory as well as at SLAC[5] (Stanford Linear Accelerator) indicated that there might be as many as five different quarks, bearing charges of 2/3 and −1/3. And, since both experimental and theoretical evidence suggested that all quarks came in groups of three, there might easily be many more types of quarks.

Alas, around this time new problems surfaced. New experiments showed apparent violations of the Pauli exclusion principle, which says that no two identical particles can occupy the same place. Three apparently identical quarks appeared to be "sitting" all in the same place. Of course, the Pauli principle was too valuable to just give up, so physicists turned their attention to trying to find some difference between the three quarks. They argued that although the three quarks had the same parity and spin, they differed in a property O. W. Greenberg of the physics department of the University of Maryland called "color." (As always, this is poetic license.)

At the moment, it can be safely said that at least six different types of quarks exist. If each then comes in three "colors," then there might be as many as 18 different quarks, though recent evidence makes this unlikely. (Further proof of some physicists' inclinations to roguishness is that many describe each of the six types of quarks as having its own "flavor.")

Finally, a theory known as quantum chromodynamics (QCD) exists to explain strong interactions between quarks. (The term "chromodynamics" derives from the fact that quarks come in "colors.") The colored quarks are then bound together by "gluons" that actually carry the force.

"Colors" are analogous to charges in the sense that the colors play the same role in QCD that electrical charges play in QED (quantum electrodynamics) except that, unlike electrical charges, there are *three* colors, while ordinary electrical charges are only plus or minus. Nonetheless, the analogy between charge and color is important in that it indicates how closely the theoretical structure of QCD resembles and is patterned after QED.

On the experimental front, by 1984 researchers had not only located considerable evidence for the W and Z particles that carry the "electroweak" force (part of the theory that explains how electromagnetic fields interact with one another), but had already found evidence for a *sixth* quark, which was dubbed "top."

One of the virtues of the quark model was that, although it greatly multiplied the number of particles, it in fact simplified and brought order to what had once been a bewildering and confusing array of new particles. This too was to a great extent the work of Gell-Mann. With the model researchers could classify and explain the action of hadrons. It now became possible to group particles into "families," each related to one another in the sense that no matter how much hadrons may differ from one another, they are all composed of quarks. Unity was emerging from confusion.

QUARKS AND NUCLEAR DISINTEGRATION

Finally, the Gell-Mann quark theory also provided a theoretical explanation of why certain particles decay faster than others. One of the features of a good scientific theory is that it can be used to make testable predictions. To the extent that the predictions are verified, the theory is confirmed. The quark model passed this test wonderfully. Using studies of the isotopes of hydrogen and helium, particularly tritium (which has a proton and two neutrons in its nucleus) and He^3 (an isotope of helium that has two protons and a neutron), scientists have found that every prediction made by the quark theory has so far come true.

Thus, the quark theory proved to be of great predictive power—all new predictions having been confirmed experimentally. Still, scientists were not completely comfortable. Where were the quarks themselves? If the quarks really were the ultimate building blocks of nature, why had none ever been produced and seen in the atom-smashing experiments in laboratories?

As of 1980, only one experiment claiming to have produced a quark seemed to have any plausibility. All other searches have been failures. These include efforts to produce them not only in the laboratory and particle accelerators, but in rocks taken from sea shells, seawater, and even the surface of the moon. One aid in this search is the reasonably well-founded speculation that quarks should be absorbed better in some substances than in others. In 1968, for instance, University of Michigan physicist David Rank[6] joined the hunt for quarks. He theorized that seawater would have absorbed quarks and they might be found in a variety of biological organisms. The reason is that given a long enough time, protons could decay into quarks, and water contains vast numbers of protons. Acting on this theory, Rank hunted for quarks in oyster shells, seaweed, plankton, and so forth, but he could find no definite evidence of them.

Another experiment devised to uncover the mysterious quark was performed at Stanford University in 1977 by William Fairbanks. He claimed to have found quarks in niobium. But although his work was treated with great respect, no one has been able to duplicate it so far.

So the quark—if it exists at all—remains in hiding. Even if it is never found, this does not necessarily spell doom for the quark theory. They could still be viewed as what philosophers of science call convenient "fictions," intended merely to make the mathematics come out right or make particle theory more consistent or complete. (For a while, many scientists looked at the neutrino in this way.) Of course, this is not very satisfying. Ultimately, scientists would like to confirm that their postulated entities really do exist as part of the "real" world.

Another approach to handling the reluctance of quarks to show themselves suggests that quarks are in fact real, but are so powerfully "imprisoned" inside their atomic shells that to "dislodge" them would require energies far greater than anyone can produce on Earth.

THE SUPERCONDUCTING SUPERCOLLIDER

The obvious answer, of course, is to keep building toys capable of even *greater* energies. One such apparatus is the SSC or "superconducting supercollider," which is scheduled to be built in Texas sometime in the 1990s, at a projected cost of $6 billion. This machine will be over 50 miles in diameter and will require an annual budget of $250 million. The energies it is expected to produce will be immense, equaling the energies found in the big bang at exactly 10^{-16} seconds after it occurred. With that kind of power, this device may allow physicists to rip quarks loose. The SSC wasn't designed specifically to track down quarks. Rather, its purpose is to probe further into the nature of matter and the forces holding it together. But it's a safe speculation that physicists will use it to hunt for quarks soon after it's plugged in.

In any case, the proliferation of particles continues. In the mid-1970s, an odd coincidence occurred: each of two teams, working independently of each other, believed that they had found a new particle. (This happens occasionally in science, as when Feynman, Schwinger, and Tomonaga all developed quantum electrodynamics simultaneously.) At this time, Burton Richter of Stanford's Linear Accelerator (SLAC) team found the new particle via experiments involving collisions of electrons and positrons. They labeled the new particle "psi."

And at the Brookhaven National Laboratory in the mid-1970s, a team headed by S.C.C. Ting of MIT found the same particle. Ting was looking for electron pairs produced by the collision of hydrogen nuclei with beryllium atoms, but the evidence also showed a new particle, which Ting labeled the "J" particle— selected because it looked like the Chinese character for "Ting."

Things got even more complicated in 1977 with the work of Leon Lederman's group at Fermilab in Chicago. His data too showed the existence of a new particle, but there was no room for it in quark theory as it existed at the time. For in quark theory, every particle was composed of some combination of quarks, and the available quarks had already been used up. That is, for a variety of reasons, the makeup and properties of the new particle could not be explained using any of the available quarks. This new "upsilon" particle, in turn, demanded another quark, which, in turn, was promptly dubbed the "b" quark and the cycle of proliferation and complications continued.

LITTLE NEUTRAL ONES

Just as quarks are the important particles involved in the strong interactions, neutrinos are critical in understanding the "weak" interactions, a manifestation of ordinary electromagnetic forces. The theory of the neutrino[7] arose from two serious problems in particle physics.

(A) The first problem occurred early in the history of the particle explosion; physicists had become worried because during beta decay, or the emission of electrons during ordinary radioactivity, the beta particle given off by nuclear disintegration was not energetic enough to match the amount of mass shed by the nucleus. In short, it looked very much as if some kind of subatomic grinch had stolen some energy sometime during the reaction.

Where was the missing energy? Wolfgang Pauli, in 1931, had an idea: perhaps another, still unidentified particle had made off with the missing energy. He further theorized that such a particle would be both massless and uncharged. (There actually was no special reason for these assumptions: they were made by most scientists, with the exception of Pauli, simply in the absence of evidence to the contrary.)

Originally, however, Pauli had proposed the new particle as a way of salvaging the idea that there were electrons in the nucleus—the theory of the nucleus extant at the time. For one thing, it was known that it was extremely difficult to see how an electron could be confined to as small a volume as a nucleus. Also, certain experimental results of beta decay did not support the idea of nuclear electrons. So far as the "missing energy" was concerned, the only real alternative explanation was that conservation of energy laws were merely statistically valid in beta decay. In other words, as historian of science Laurie Brown suggests in an excellent article, "The Idea of the Neutrino," it is often overlooked that Pauli originally supposed that the neutrino was part of the nucleus, "with a small but not zero mass, together with the protons and the electrons."[8] According to Pauli, these would be confined to the nucleus by magnetism. Hopefully, that would solve the problems of the missing energy in beta decay.

(B) The second problem was this: According to the law of conservation of spin, it was known that a neutron can and does decay into an electron and a proton. All of these have spin 1/2. But then one is committed to saying that a neutron (spin 1/2) can somehow give rise to an electron (spin 1/2) as well as a proton (spin 1/2). Where does the *extra* spin 1/2 (say, of the electron) come from? Well, if one assumes that the proton has a *positive* spin of 1/2, and the electron has a *negative* spin of 1/2, then the latter two spins cancel each other out and so add up to zero. Now all you have to do is assume that a neutrino, also with spin 1/2, is given off along with the electron and the proton. That is, *three* particles with spin 1/2 are given off, rather than two—the proton, electron, and neutrino. Since the spins of the electron and proton cancel each other, the net result, in effect, is that a neutron of spin 1/2 has given rise to a neutrino of spin 1/2 and the equation balances.

To many, this seemed like a typical dodge: What do you do when you confront a new problem? Invent a new particle to solve the problem.

Nevertheless, many physicists were convinced that the neutrino did exist. Most important of these was the great Italian genius, Nobel laureate Enrico Fermi. When first hit with the idea at the great physics conference in Rome of 1931, Fermi, after numerous talks both with Pauli and with other theorists, liked the concept. Niels Bohr, however, did not.[9] He preferred to believe that ordinary conservation laws simply did not apply on a nuclear scale. And there was to be support for Bohr's approach. As Yang and Lee would show in the 1950s, for example, parity wasn't conserved under certain conditions.

Fermi, however, persisted, calling the new particle a "neutrino," Italian for "little neutral thing." According to Laurie Brown, not only was the concept critically important for the theory of the nucleus, but some, like Robert Oppenheimer, believed that neutrinos (or "neutrons" as they were then called, before Chadwick's discovery) would be useful in explaining cosmic rays.[10] But only a year later, their researches showed no confirmation and they abandoned the idea.

Although the neutrino was both aesthetically satisfying and did patch up the holes in theory created by experiment, physicists, as always, would have been happier to actually *find* the things. But how? Physicists generally succeed in identifying particles by virtue of their unique properties. Thus, for example, Anderson identified or "caught" the positron because of its unique and unexpected charge and mass (its mass was the same as an electron's, but it had a positive charge).

But neutrinos at first appeared to have virtually *no* properties. (Not merely no distinguishing properties, but quite literally, no properties at all, illogical as that may sound.) They did not show electromagnetic force, had little, if any, mass, and were uncharged, as their name implies. Because of these odd features, they could pass through a shield of any substance whatsoever no matter how unimaginably thick the shield was.

But, in 1953, a miracle occurred. Clyde Cowan and Fred Reines at Los Alamos set out to track down the elusive "neutrino" at a site near the huge nuclear reactor on the Savannah River in Georgia—a logical step, since neutrinos can be expected in virtually any nuclear reaction.

Their idea had a new twist as well. They wanted to allow any antineutrinos (really just the name given to an ordinary neutrino when it is emitted along with an antielectron, or positron) to bombard hydrogen nuclei (protons) in large water tanks, in the hope that they could detect a proton "grabbing" an antineutrino. That, in turn, should lead to a proton being converted to a neutron, which, in turn, should give off a *positron*.

The beauty of the process was that both positrons and neutrons can easily be identified. Positrons would annihilate an ordinary electron in about a millionth of a second, producing easily detectable gamma rays in the process. The scintillating devices would then record the passage of photons. The gamma rays would thereby provide conclusive proof of the existence of antineutrinos or, as they are now called, "neutrinos." With this method, at long last, science caught the little monster.

THE WEAK FORCES

At this point in particle physics, scientists were becoming even more eager to establish "unity" in their conceptual schemes. That is, they were trying to show that the various different forces discussed so far were really just different manifestations of a single force. Many believed that this single, basic force could appear "in disguise," as it were. It was known already that both electromagnetic and "strong" forces took place via an exchange of particles. The electromagnetic force was generated by the exchange of photons, while the strong force appeared when Yukawa's pi meson was exchanged within the nucleus.

Could it be that the *weak* intranuclear forces were similarly mediated by particle exchange? Further, could the "weak" and electromagnetic forces be unified? On the basis of what was known about the weak forces, scientists could predict what property this yet undiscovered particle should have. Since Indian physicist S. N. Bose, along with Einstein, had made a thorough study of bosons, this new particle was named the "intermediate vector boson."[11]

WEINBERG AND GLASHOW

A further insight into the nature of the weak forces was achieved by physicists Steven Weinberg and Sheldon Glashow.

Weinberg received his B. S. from Cornell in 1954, earning a doctorate from Princeton in 1957. Early on he was interested in Feynman's work and, importantly, broken symmetry. The latter is the idea that at the very beginning of the universe, all particles and forces were identical, but that within fractions of a second, these identities or "symmetries" were broken. (For example, the electromagnetic force a brief instant after the big bang was no longer identical to the force of gravity.)

Weinberg was also interested in scattering theory and particle physics. Glashow earned his B. S. at Cornell in 1954 and his doctorate from Harvard in 1959, working under Julian Schwinger. He focused mainly on the unification of forces in physics. The work for which he, Salam, and Weinberg received the Nobel Prize unified electromagnetism and the "weak" forces in the atomic nucleus.

In the theory Weinberg proposed in 1967, working with Glashow, they utilized the idea of "symmetry breaking" in a theory known as "gauge" theory. All of this is an important link between particle physics and cosmology. Scientists believe that when energies are enormously large, there can be no difference between electromagnetic force and the weak forces. (And at even higher energies, such as existed in the first moment after the Big Bang, *all four* forces were indistinguishable from each other.) But, as the cosmos cooled, gradually the four forces split off from one another. Weinberg and Glashow called this "symmetry breaking," since the original equivalence of "symmetry" of the four forces vanished with dropping temperatures.

Of course, none of this has really been "proved"—indeed, "proof" is almost an inappropriate concept in the highly speculative realms of theoretical astrophysics these days. Still, the 1965 discovery of microwave radiation that was produced by the big bang is strong evidence for such speculations. Also, the lack of abundance of certain elements (e.g., hydrogen and helium) in the universe today allows scientists to make reasonable guesses about what was going on at the moment of the big bang.

That, of course, happened with electromagnetism and the weak forces, which also split off from each other with dropping temperatures. Weinberg and Glashow's brilliant suggestion was therefore a proof that these two forces could be treated as merely different aspects of one and the same force. In doing this, they accomplished part of Einstein's dream of a "grand unified theory" or GUT—the ultimate unification of all forces in nature. As such, their Nobel Prize was one of the most deserved in history.

In their 1967 paper on all of this, they proposed that after the breaking of symmetry between the forces, the photon would remain massless and carry the electromagnetic force. They also suggested that there would be three other particles: the W^+, W^-, and Z^0 particles. These would remain with some finite mass, about 90 times the mass of a neutron. The heavy, short-range W and Z particles would carry the weak force, and Weinberg and Glashow predicted the existence of a "weak neutral current." In this conception, at extremely high energies, the electromagnetic and weak forces cannot be distinguished from one another. (Unification!)

Also, in a radical departure from conventional thinking in particle physics, Weinberg suggested that the Z particle would be a force-carrying particle that is *uncharged*—a phenomenon previously thought impossible. The canonical view was that such a particle must have either a negative or a positive charge.

Subsequently, in 1973, their views were confirmed by the discovery of "weak neutral currents" (which had been predicted by Glashow and Weinberg) by scientists at the Fermi National Accelerator Laboratory in Chicago as well as the European particle laboratory, CERN (Consile Europeen pour la Recherche), in Switzerland.

Still more complications came when, in 1975, a Stanford group found a particle connected with the weak interactions that was far heavier (1.8 GeV) than any previously discovered. Indeed, the number of particles responsible for weak interactions appears to be increasing just as fast as the number of quarks. As more powerful atomic accelerators are produced, it is not unreasonable to suppose that even more and perhaps even heavier weak interaction particles will appear. As that happens, of course, the dream of reducing the building blocks of nature will recede even further into the distance. But there is hope: the proliferation of particles may have an end, and this has to do with complex issues related to the big bang. If this does not work out, it is just conceivable that there may be no end to the complexity of the universe.

CHAPTER 18

Richard Feynman and Quantum Electrodynamics

As one observer described a movie showing Feynman at work, ". . . the guy doing the talking seemed peculiar—not an oracle but a *shvitzer*, with a voice and manner disconcertingly like Art Carney's. The movies had titles like 'The Law of Gravitation' and 'The Relation of Mathematics to Physics,' and the *shvitzer* with the chalk was the American physicist Richard Feynman. . . ."

THE MAN: AN OVERVIEW

The scientific world credits Richard Feynman, the quixotic, mysterious scientist and bongo-drum virtuoso, with almost single-handedly developing the science of quantum electrodynamics (QED), which explains one of the four forces in the nucleus—the electromagnetic force. In his Nobel Prize acceptance speech in 1965, he speculated about whether "the theory is simply a way to sweep the difficulties under the rug," referring to the fact that earlier attempts to develop QED involved infinite quantities—the infinite usually being a sign in physics that something has gone wrong. Feynman, though capable of consummate arrogance, was uncharacteristically modest on this occasion. Yet such skepticism is vital to science: the theorist who believes he has reached absolute truth will soon encounter a reality shock, as history has proven time and again. The scientist must always be ready to abandon a theoretical idea when circumstances or experimental evidence warrant that he or she do so. As Feynman said on another occasion, "It is of great value to realize that we do not know the answers to different questions. This attitude of mind— this attitude of uncertainty—is vital to the scientist, and it is this attitude of mind which the student must first acquire."[1]

Besides his work on "renormalization" (a mathematical technique for eliminating some annoying results in QED) to remove discrepancies between older work in QED and some of his own newer experimental results, Feynman also contributed to the scientific understanding of radioactive decay and the structure and forces within the atomic nucleus.

EARLY LIFE

He was born in Brooklyn on May 11, 1918. His father was a sales manager of a uniform factory. When his wife was pregnant, he predicted that the baby, if a boy, would become a scientist.

In a *New York Times* interview in October 1967, Feynman said that his father had taught him "continuity and harmony in the world. He didn't know anything exactly, whether the insect had eight legs or a hundred legs, but he understood everything. And I was interested because there was always this kick at the end—a revelation of how wonderful nature is."

He graduated from Far Rockaway High School in 1935 and enrolled in MIT. He received his B.S. with honors in 1939 and went on to Princeton for graduate work, writing his dissertation with the fabled John Archibald Wheeler. Wheeler was one of the greats in physics who had worked extensively in the field of atomic energy.

Because war was on the horizon, much of the scientific community began to turn its attention to military work. Feynman was no exception and during this period of military frenzy, he worked at the Frankford Arsenal in Philadelphia, where he designed a computer geared to aiming artillery. At the same time, he continued his graduate work at Princeton. Though slowed somewhat by the Frankford work, he received his doctorate in 1942.

THE MANHATTAN PROJECT

Inevitably, Feynman became involved in the Manhattan Project, which gave the United States victory in the atom bomb race before Germany could complete its own atomic research. Almost immediately after he joined the project, tragedy marred his work. Physicians diagnosed his first wife, Arlene Greenbaum, as terminally ill, and she died in 1945. His second marriage ended quickly in divorce. In 1960, Feynman married his third wife, Gweneth, whom he met in Switzerland, producing two children, Carl and Michelle.

Starting his work at Princeton University, he remained in Los Alamos working on the Manhattan Project through the war years, meeting many of the world's elite in science, including Teller, Bethe, Fermi, and Oppenheimer. Oppenheimer, in particular, impressed Feynman. He had contributed a respectable amount to theoretical physics, principally in the form of cosmic ray studies and quantum mechanics.

Never known for taking anything too seriously, Feynman acquired a reputation as a prankster by picking top-secret security locks and leaving notes signed "Feynman the safecracker."

As a member of the Project, Feynman was among those who witnessed the first experimental explosion at Alamogordo, New Mexico, in July 1945. Although he was happy to have contributed to the Project, his exhilaration soon turned to *angst*. As he said in his memoirs entitled *Surely You're Joking, Mr. Feynman*, "I can't understand it anymore . . . but I felt strangely then. . . . I would go along and I would see people building a bridge . . . and I thought, they're crazy, they're crazy, they just don't understand. . . . Why are they making new things? It's so useless. But, fortunately, it's been useless for almost forty years now, hasn't it?"

After the war, with the encouragement of Hans Bethe, he accepted the post of associate professor of physics at Cornell University by Lake Cayuga in Ithaca, New York. His office in Rockefeller Hall overlooked the beautiful arts quadrangle, with its mesmerizing statues of Andrew Dickson White and Ezra Cornell.

QUANTUM ELECTRODYNAMICS

Feynman's greatest contribution, of course, was in the field of QED, which he all but invented. However, the great Max Planck had anticipated some of the ideas in this area as early as the 1890s.[2] But Feynman had not yet been born and the time was not yet ripe for QED. Although the roots of the new field also appeared in some theoretical work as early as the 1920s, Feynman was the very first to put these ideas into an intuitive, pictorial form, making them more readily understandable.

THE POSTWAR PERIOD IN PHYSICS: LAMB'S WORK

To understand QED, it is necessary to first look at some of the historical scientific work that inspired QED. A major event in Feynman's career, as well as in the careers of co-founders of QED like Julian Schwinger and Freeman Dyson, which soon brought their ideas to the full attention of the scientific community, was a series of conferences held in the United States after World War II. Its participants included, among many others, such greats as Rabi, Pais, Schwinger, Feynman, Wheeler, Teller, and Weisskopf. Physicists and the National Academy of Sciences held the conferences shortly after the war to discuss recent work in theoretical physics. The first meeting was held at Shelter Island, New York; the others were at Pocono Manor and Oldstone.

The conference at Shelter Island was given a special appeal by the fact that Feynman, Linus Pauling, Edward Teller, Robert Marshak, and Abraham Pais were present, among many others— scientists who had already changed the landscape of physical science. The Shelter Island conference on quantum mechanics opened on June 2, 1947.

From a human interest viewpoint, the conferences themselves were a study in scientific egoism—to the point of absurdity. Notorious prima donnas of science, Pauling in particular, caused the organizers considerable annoyance. According to one of the participants, MacInnes, some, including Pauling, "have behaved like prima donnas at least, and are apparently badly spoiled. . . . All that remains to be done . . . is to see that enough transportation is provided to get the men there and back, with the addition of two ladies, Mrs. Pauling and Mrs. Von Neumann, both of which have been added by means of polite blackmail. . . ."[3] Nevertheless, the conferences took important steps forward in solving the problems of QED.

SCIENTIFIC WORK AT THE CONFERENCES

Among the scientific accomplishments of the Shelter Island conference in particular was that it forced the physics community to rivet their attention on some difficulties for QED created by a series of experiments. First there were the experiments that American physicist Willis Lamb had conducted, the results of which contradicted the Dirac equation and the Dirac relativistic theory of the electron. P. A. M. Dirac in 1928 had published his seminal paper on the electron which described all of its properties as they were then known, including spin, charge, and so forth.[4] According to his theory, there were two different orbitals, the s and p orbitals, which an electron of hydrogen could occupy. In an atom of hydrogen, a single electron orbits the nucleus. But nature allows the electron two possible orbitals or "paths" around the nucleus. To drop from the higher orbit to the lower one, the atom must give off a photon, while to move to the higher orbit, the atom must absorb a photon. And because the photon absorbed has to correspond exactly to the difference in levels between orbits, these types of transitions yield spectra of very clearly spaced lines. According to Dirac, the amount of energy in these two orbitals should be the same. After working at Columbia with I. I. Rabi, Lamb became interested in the so-called "metastable" (*unusually* stable) states of hydrogen. Under ordinary conditions, an electron occupying a higher or "excited" orbital around the nucleus, such as is the case when an atom absorbs electromagnetic radiation or when it collides with another particle, will quickly drop to a lower orbit and emit a photon while doing so. But in comparison to the ordinary stability electrons usually display, the metastable state lasts far longer. Electrons in these metastable states, for various theoretical reasons, can drop to a lower state only by emitting *two* photons at once. But since the latter is a far less likely occurrence than the ejection of one photon, the electron remains in the metastable state much longer.[5]

It was in his work with Robert Retherford at Columbia, however, that Lamb's experimental work led to a historic moment in the development of QED. He and Retherford found that there were far more lines in the hydrogen spectrum than anyone had ever seen before. Lamb and Retherford discovered that ordinary spectral lines of hydrogen actually consist of a series of even *greater* numbers of lines extremely close to one another. Because very close study and high resolutions are needed to see these "extra" lines, these had never before turned up in experiments.

But Lamb, in finding the two "sublines" within the normal spectral line, showed that the two orbitals did *not* have identical energy levels. Dirac was wrong. And since QED depended heavily on Dirac's work, this put the validity and usefulness of QED in serious jeopardy. Physicists had previously supposed that QED was a valuable tool for describing the properties of atomic systems, at least the simpler ones—until the work of Lamb and some of his colleagues. The Lamb shift, and the need to repair Dirac's theory, captivated the interest of the physics community, leading ultimately to the great postwar work in QED.

By May 1947, Weisskopf, Schwinger, and others on their way to New York to attend the Shelter Island conference were worried about Lamb's results, contradicting Dirac's predictions. They believed it was essential to solve them if QED were ever to work. At the time, neither Feynman nor anyone else had worked out a completely satisfactory solution to the difficulties. Thus, they too began fresh assaults. Eventually, it would be this "Lamb shift" that led to Feynman, Tomonaga, and Schwinger's work in "repairing" Dirac's theory of the electron. By taking into account what Dirac did not know, these theorists were able to predict and explain the Lamb shift. (It should be noted that we are not talking about three "different" theories in saying that QED was the result of these three scientists: Tomonaga, Feynman, and Schwinger ultimately said the same thing, but in different ways and from different perspectives.)[6]

To work out the snags, Feynman devised rules for bringing the theory of QED in line with experimental laboratory results. As Feynman later unabashedly called it in his book *Quantum Electrodynamics: The Strange Theory of Light and Matter*, it was the ". . . jewel of physics." Despite the formidable name, the ideas are relatively straightforward. Science had known for eons that electromagnetism was a basic force in nature. But no one had produced a satisfactory explanation of it. Feynman's idea involved two important concepts:

a. Particles transmit the electromagnetic force.
b. Particles can literally appear out of nothing, disappearing again as soon as the force has been transmitted. (Physicists call such particles "virtual" particles.)

The Heisenberg "uncertainty" relation, which says that it is impossible to know with certainty what the energy level and the duration of the existence of a particle are, sanctions principle (b). Because of the theoretical uncertainty of energy (which is, after all, just a form of mass, according to Einstein's famous $E=mc^2$), it is possible for a particle to "borrow" enough energy to exist for a while and then vanish again. Feynman diagrams describe these interactions between particles.

Principle (a) is harder to understand. Think of two skaters tossing a ball back and forth. The exchange of the ball causes a "recoil" effect, driving the skaters farther apart. The particle exchanges that transmit the electromagnetic (and other) forces are sort of the *reverse* of this. For example, imagine two skaters standing back-to-back with specially made *boomerang* balls. One skater then throws a ball forward, making him recoil *backward* toward the other skater. His ball, being a boomerang, reverses direction and heads for the skater standing behind the first one. The second skater catches the ball, throws it *forward*, it boomerangs backward, and the second skater also "recoils" backward, toward the first skater. By continually tossing the ball in this way, the recoil gradually pushes the two skaters toward one another. In much the same way, the particles in the nucleus "toss" virtual particles to one another and drive themselves (i.e., the nuclear particles) closer to one another.

THE PLAGUE OF INFINITY

There was, however, a huge problem in QED—the notorious problem of the "infinities" as scientists referred to it. Some of the calculations were yielding "infinite" results, or particles that were infinitely heavy. It is a very general principie in science, philosophy, and math that any time one runs into any kind of "infinite" number or quantity, theorists regard it as a sign that something has gone badly wrong somewhere. Virtually no scientist, philosopher, or mathematician believes that any quantity can actually be "infinitely" large. Physicist H. A. Kramers described these problems in a paper in *Nuovo Cimento*, 1938, as follows:

> At almost every important stage of the development of quantum mechanics, not only were new positive results added to what had already been achieved, but also certain "defects" revealed themselves. . . . When in 1926 the field of free Maxwell radiation was quantized, the infinite zero-point energy revealed itself. . . .[7]

His main point was that the infinite shift in spectral lines (lines representing various wavelengths of light) was intimately related to the mass of the electron considered as a singularity. (A "singularity" would be what is left if some unfathomable force managed to compress a given amount of matter into a mathematical point. The electron is the "zero-point" Kramers refers to, and its energy, under the usual calculations, turned out to be infinite. It was precisely this problem that Feynman intended his version of QED to remedy.

FEYNMAN DIAGRAMS

As the noted Princeton physicist Freeman Dyson said in the October 29, 1965 issue of *Science* magazine, Feynman essentially reinterpreted "almost the whole of quantum mechanics and electrodynamics from his own point of view." A good example are the fabled "Feynman diagrams," which allowed one to visualize and easily calculate the interactions between photons and electrons. (For years Feynman had the diagrams drawn on his van in Pasadena.) They were pictorial representations of the mathematical content of theories, insights that Dyson would later expand upon to show how they could be used to clarify problems in other areas of physics.

Ever since his Ph.D. work in 1942, Feynman had been able to put the mathematical ideas of quantum mechanics—at this point little more than nonintuitive abstractions—into intuitively plausible pictorial representations of the various paths a particle could take. By making the ideas of quantum mechanics visualizable in this way, he made quantum mechanics as well as quantum electrodynamics intuitively plausible. The scientist could now visualize what had been previously describable only in abstract mathematics.

In the diagrams Feynman symbolized probabilities, for instance, by drawn *arrows*. The length of each arrow squared, for example, indicates the probability that a given physical event will in fact occur. The longer the arrow, the more likely the event is to occur. The arrows can cancel each other completely or in part, or reinforce one another completely or in part, depending on the probabilities. Every vertex on the diagrams encapsulates a set of mathematical figures (integrals) that essentially show how the particles have behaved over a given period of time. Moreover, the arrows and diagrams allow the scientist to picture three other fundamental actions of quantum electrodynamics:

1. A photon moving from place to place
2. An electron moving from place to place
3. An electron absorbing or giving off a photon

Again, we see the previously mentioned idea that exchanges of *particles* carry these forces. Here, two electrons will repel one another because they "throw" (virtual) photons back and forth, like our imaginary skaters on the skating rink. The diagrams describing each of these possible events in QED provided an easy and convenient way of keeping track of the related mathematical expressions. These proved just as useful to physics as blueprints are to engineers, thus making QED one of the most powerful fields in physics, despite the absurd "infinities" and other logical inconsistencies that continued to plague the field. They helped physicists calculate the interactions between all kinds of sub-atomic particles, including quarks, gluons, bosons, and so forth—an application that extended way beyond even Feynman's own original purposes. Moreover, physicists have verified via careful experiments all predictions based on QED.

TRAVEL IN QED

With QED, a scientist could solve almost any problem in electrodynamics. This, in turn, led to the discovery of still more insights and breakthroughs. Feynman caused considerable commotion, for example, when he suggested that an electron, after ejecting a photon, could make up for this by going *backward* in time to absorb another photon.

This is based on the "uncertainty principle" described earlier. What happens is that an electron emits a photon and another electron then absorbs the photon. Because of the uncertainty principle, it is impossible to know both the duration of the existence of a particle and its energy exactly. As described earlier, the principle allows that particles such as photons can pop into existence for brief periods. Nature literally creates them out of nothing, via the uncertainties in energy. As with all particles created in this way, they do not exist permanently and scientists therefore call them "virtual" particles.

Only after capturing the second photon would the electron continue forward in time. Feynman once quipped, trying to make the point of the diagrams clear, "You know how it is with daylight saving time? Well, physics has a dozen kinds of daylight saving time." Finally, the existence of such particles is beyond question, for even though they cannot be seen, their effects are real.

BOHR'S RESISTANCE TO THE MYSTERIOUS FEYNMAN CONCEPTS

Feynman presented his ideas at another of the postwar conferences, the Pocono conference in April 1948, at Pocono Manor (a resort area between Scranton, Pennsylvania, and the Delaware Water Gap). Not everyone liked what Feynman was doing. For one thing, his presentations were arrogant, breezy, and "off the top of his head." Nor did it help his cause that he presented his approach to QED the day after Julian Schwinger (now professor of physics at UCLA and co-winner with Feynman of the Nobel Prize) had carefully presented his more detailed, systematic, and extremely complicated mathematical treatment of QED.[8] (Shin'ichiro Tomonaga of Tokyo, in contrast to both Feynman and Schwinger, presented a simple, conservative explanation of QED, using little more than the quantum mechanics of the 1930s.)

Bohr, in particular, was growing increasingly irritated with the young "schvitzer." Bohr found particularly distasteful the bizarre idea that electrons could go backward in time. Also, Bohr believed, plausibly enough, that the diagrams violated Heisenberg's uncertainty principle. Since the German physicist had proved convincingly that it was impossible in principle to measure both the position and the velocity of electrons, how could Feynman presume to claim a particle could follow precisely definite paths through space?

And Bohr doubtless felt particularly well equipped to express himself. He had a long familiarity with QED, having worked with Rosenfeld in 1937 at Princeton on the problem of measurement in QED, even giving several lectures on it. John Archibald Wheeler, also a participant at the Pocono conference, described Bohr's reaction to these problems in his unpublished essay "Conference on Physics—Pocono Manor, Pennsylvania, 30 March–1 April 1948," as follows:

> [he believed] . . . Schwinger's treatment is an advance in bringing the theory [quantum electrodynamics] into a regular order whose effects can be interpreted. As regards the infinities he [Bohr] had other views than those just discussed. He objected to Feynman's view of the electron going backward in time. . . .[9]

The reaction of so great a figure as Bohr probably depressed Feynman, although his natural self-confidence won the day. Not long after Bohr's stinging rebuke, Feynman even published his epochal papers, "Space-Time Approach to Quantum Electrodynamics" and "The Theory of Positrons." Feynman first published these articles in 1949 in *The Physical Review*.

Despite the objections of Bohr and others, Feynman's work—the diagrams and the related "Feynman integrals"—quickly became standard apparatus for understanding the behavior of subatomic particles. In its "renormalized" form, QED now had enough power and consistency to explain most of the experimental results, the shifts in spectral lines of hydrogen demonstrated by

Lamb, as well as other properties of electrons in atoms, thus correcting Dirac's original theory.

THE SCHWINGER AND TOMONAGA APPROACHES TO QED

Feynman was not the only one working on QED. (Again, the three versions of QED are merely different ways of expressing the same underlying theory: all are equivalent to one another, just as one can express or picture the same data by a bar graph or a line graph.) As a result of the great publicity of the conferences, both Tomonaga of Tokyo University and Julian Schwinger of Harvard presented solutions to the problems Feynman had worked on, although their solutions were nonpictorial and intuitively less appealing. Nevertheless, the scientific community judged both to be fully the equal of Feynman's in working out the ideas of QED.

Soon after the Pocono conference, its unofficial mentor, Robert Oppenheimer, received a letter from Tomonaga, accompanied by his wartime article on QED. Oddly, the Japanese physicist had only become interested in the problems bewitching QED because of an article that appeared in *Newsweek* in 1947, describing the Lamb experiments. Soon after that, Oppenheimer, putting all possible wartime hostilities aside, made certain that everyone knew about Tomonaga's important work. Oppenheimer's move was objective science at its best. He sent copies of Tomonaga's correspondence to everyone at the Pocono conference.

Schwinger, building on the early work of theorists like Podolsky and Dirac, offered a massively complicated, systematic, and difficult-to-follow account based on a series of mathematical manipulations of the state of a single electron. Tomonaga also intended his version of QED to explain and clear up the discrepancies and the problems of the infinities resulting from the same experiments that inspired Feynman's work.

OTHER BENEFITS FROM THE CONFERENCES

It is worth emphasizing other important features of the conferences. Through the efforts of men like Duncan MacInnes of the Rockefeller Institute as well as Karl Darrow of the American Physical Society and Frank Jewett of the National Academy of Sciences, these postwar conferences allowed the best and brightest of the American physicists to meet for the first time to discuss pure science without having to worry about finding better ways to destroy people.

In recognition of their work, Feynman, Tomonaga, and Schwinger shared the 1965 Nobel Prize in physics.

LATER WORK ON THE SUPERFLUIDITY OF LIQUID HELIUM

By the early 1950s, Feynman had turned his attention to other areas of physics, offering, among other things, a quantum mechanical theory of liquid helium. Russian physicist Lev Landau had devised the theory back in the 1930s. In attempting to explain some of liquid helium's bizarre properties, he postulated a type of disturbance called a "phonon," or a sound wave of extremely high frequency. According to Landau's theory, the motion of the atoms in a solid generated such waves. The waves, in turn, caused the observable bizarre behavior of helium. Thus, at temperatures approaching absolute zero, liquid helium displayed superfluidity, an immensely low viscosity that caused it to flow *upwards*.

After studying Landau's work, Feynman confirmed the existence of phonons, even adding to their physical description. He suggested that liquid helium at such low temperatures enters what physicists today call a "phase change," a concept known about since the time of America's first theoretical physicist, J. Willard Gibbs. In this state, its molecules drop to extremely low energy levels. At that point, they begin to obey a weird set of quantum principles known as Bose–Einstein statistics. Physicists confirmed all of this through careful experiments in 1958.

THE WEAK FORCES IN THE NUCLEUS

From this work, Feynman turned to the so-called "weak force"—one of four principal forces within the nucleus of the atom. (The other intranuclear forces were the electromagnetic force, gravitation, and the "strong force.") Evidence of the weakness of the "weak force" is the fact that the nuclei of some atoms slowly and naturally fall apart in their special kind of radioactive decay, commonly known today as "beta decay." Scientists had observed such evictions of electrons from the nucleus for many years, but the mechanism by which this occurred had always remained obscure.

By the 1950s, scientists were becoming increasingly uneasy of the fact that radioactive decay, though common in nature, had remained an enigma for so long. One reason for the heightened concern was that recent theory had postulated two types of beta particles—tau and theta. What was odd in the extreme was that the two particles appeared to be in all respects identical, except that they appeared to violate the law of conservation of parity, one of the many "conservation" laws long held to be canonical. According to this principle, nature makes no distinction between left and right. (The laws of physics will operate equally well in the real world, or in Alice's "looking-glass" world.)

Feynman along with others decided to make a frontal assault on this issue. He soon began to suspect that violations of parity were, in fact, a property of the weak interactions—speculations they discussed at a conference in Rochester, New York, in 1956. Two other physicists, T.-D. Lee of Columbia and C. N. Yang of the State University of New York at Stony Brook, also began to suspect the same thing. The Columbia University physics department, on January 15, 1957, reported that physicist Chien-Hsiung Wu had confirmed all of these musings in an experiment performed at

Columbia. This, of course, was the fantastic discovery of "parity" violations which overthrew a principle long believed to be absolutely inviolable and which earned Yang and Lee the Nobel Prize. As a consequence, science now understands beta decay very well and knows it to be one of the "weak interactions."

GELL-MANN AND STRANGENESS

Though Feynman has received the lion's share of attention in recent years, the work and career of Murray Gell-Mann of Caltech is also important. Less flamboyant and more conservative in manner than Feynman, he was responsible for many important breakthroughs in what is now called "particle" physics. Gell-Mann grew up in New York City and received a bachelor's degree from Yale in 1948. He then went on to MIT, receiving a Ph.D. in physics in 1951.

One of his earliest accomplishments was his study of the quality of "strangeness."[10] He suggested the term "strangeness" because the unusually slow "half-life"—the time needed for half of a given amount of a substance to decay—of some substances violated the then-current theory of particle behavior. The chief difficulty in explaining this strange behavior was the fact that, on the one hand, the speed at which nature created the "strange" particles in collisions with other particles suggested that the strong and very fast-acting nuclear force was controlling them. But, their unusually long half-life suggested that it was the weak nuclear force that orchestrated their behavior. *Which* force was really controlling them? What was going on?

Soon he realized that "strangeness" was conserved in particle interactions, just as physicists believed that energy, parity, and time would be conserved. That is, just as the amount of energy coming out of a reaction must equal the amount going in, so the degree of "strangeness" of a substance must be the same coming out as going in. Gell-Mann's logic was relentless: If the strong forces were guiding strange particles, then strangeness would be conserved. But experiments proved that strangeness was not conserved. Therefore, the strange particles had to be controlled by the weak forces.

It was only after this work that he began his collaboration with Feynman on the weak forces. They published their results in a paper titled "Theory of the Fermi Interaction" in *The Physical Review* in 1958.

During the 1960s, Feynman studied and tried to make sense out of a quantum theory of gravity. Unsatisfied, he soon returned to particle physics in 1968. At the time, physicists were working on the phenomenon of "scaling" which appeared to offer promise for understanding the "strong forces"—the forces that hold the nucleus together. The phenomenon might, they theorized, reveal even more basic particles than the proton—perhaps the long-sought-after *ultimate* building blocks of matter.

In pursuit of this quest, Feynman visited the Stanford Linear Accelerator Center in California in 1968 and began constructing a theory that explained the results of protons being assaulted by other high-energy electrons. He speculated that protons could be made up of particles he called "partons." He next suggested some ideas for experiments that might confirm their existence. Commenting on his genius, physicist Henry Kendall said, "Feynman being Feynman, if Feynman says that you fellows are observing pointlike constituents in the nucleus, then you pay attention."

As it turned out, Feynman's concepts were almost identical to some proposed by Gell-Mann. Feynman's "partons" were essentially the same as Gell-Mann's "quarks." Gell-Mann had arrived at the idea of quarks while at MIT, working on classifying the "strange" particles into "multiplets," or groups of particles sharing a certain quantum property while differing in other properties. (A group of people all having red hair, but different weights, ages, heights, etc., would be a kind of "multiplet.") Feynman published the results of his investigation of "scaling" in 1970. By the early 1970s, the quark–parton theory had established itself in the physical community. In essence, this model created the theory of quantum chromodynamics, currently the most popular theory attempting to explain what holds the nucleus together.

THE CHALLENGER DISASTER

In the mid-1980s, one of the centuries truly great catastrophes struck NASA and the world. The Challenger rocket exploded in a huge cloud of white smoke, keeping Americans from coast to coast glued to their television screens. The death of Christa McAuliffe particularly, the schoolteacher-astronaut, threw the nation into shock.

During the commission hearings, scientists theorized that a defective neoprene "O-ring," part of the mechanism that seals in the volatile fuel, might have caused the tragedy.

After hearing a NASA spokesman pontificate on how "hard" it was to test such theories, Feynman merely asked for a glass of ice water, dropped a piece of the O-ring into it, and proved that the low temperature caused it to become brittle and inflexible. In that state, it was no longer able to expand and contract to keep the fuel joints sealed. In scientific terms, a "phase change" had occurred in the O-ring. He issued his findings separately in a report titled "Personal Observations on the Reliability of the Shuttle," in the September 1986 issue of *Esquire*. Science journalist Michael Ryan commented that Feynman "took NASA to task . . . for its overblown claims, its underdone management" and added a "rare and plaintive grace note about the courage of Christa McAuliffe. . . ."

Beyond his strictly scientific work, he has published his memoirs under the title *Surely You're Joking, Mr. Feynman!* The great Princeton physicist Freeman Dyson says of Feynman's book that it is a "wise, funny, simple, profound, cool, passionate and totally honest self-portrait by one of the great men of our age." In addition to the Nobel Prize, Feynman has captured innumerable other awards, including the Albert Einstein Award in 1954, the Niels Bohr Gold Medal in 1973, and the Oersted Medal in 1971.

Richard Feynman died of cancer in 1988.

CHAPTER 19

Glashow, Weinberg, and the Search for a Unified Field Theory

THE QUEST FOR UNITY

For centuries this search has been the Holy Grail of philosophers, theologians, and scientists. Why this should be so has never been completely clear. As philosopher Ludwig Wittgenstein said in the *Blue Book*, the differences between things can be as revealing as the similarities. Nevertheless, the idea of unity fascinates: perhaps it is because of the simplicity that unity implies.

The ancient Taoists believed that every happening in the universe could be explained by the interaction of just two forces, yin and yang, yin being the subordinate or "feminine" principle and yang the dominant or "masculine" principle. Much later, the ancient Greek pre-Socratic philosophers such as Thales, Anaximenes, Heraclitus, and others also sought to explain the workings of the universe with as few principles as possible. Thus, Thales believed water was the dominant principle, while Anaximenes argued vigorously that it was air, and so forth.

By the time the physicists of the 19th century had begun their own speculations, nature was still a disparate lot. There were at least three seemingly independent forces operating in the cosmos—electricity, gravity, and magnetism. As had their predecessors, these scientists wished to reduce the number of explanatory principles to the smallest number possible. In the 1860s, the great British physicist James Clerk Maxwell took a giant step in this direction when he recognized that there was such a thing as an "electromagnetic" field, thereby combining electricity and magnetism. From this basis, Maxwell's work allowed scientists to further understand the nature of light and radio waves.

Still, with the development of relativity theory and quantum mechanics in the 20th century, physicists began to appreciate that there were still more forces waiting to be "unified." With the progress of nuclear physics, scientists realized that there was a force holding the nucleus together, which they called the "strong" force—so named because these forces had to be powerful enough to overcome the repulsive energy of the particles within the nucleus. But the forces kept proliferating. Soon, physicists had to deal with the electromagnetic force and the "weak" force (responsible for radiation).

In 1946, physicists Abraham Pais and C. Moller classified all particles that did not feel the strong force as "leptons." Each force had its own unique properties. The forces of electromagnetism (carried by photons, which we can detect as radio waves, visible light, or infrared radiation) and gravity (carried by gravitons) had a range of infinity, since photons and gravitons had zero mass. (Physicists know from quantum mechanics that the range of force of a particle varies inversely with its mass.)

Of all of the forces, the strong force is the most powerful, as its name implies. Next is the considerably weaker electromagnetic force, which keeps the electrons in orbit around the nucleus of an atom. Weaker still is the "weak" force, which explains the phenomenon of radioactivity, and weakest of all is gravity.

Then, in 1962, physicist L. B. Okun suggested the term "hadron" as a collective word for all of the subatomic particles that feel the strong force, such as protons and neutrons. Yet the new "strong" forces in the nucleus had an extraordinarily limited range—no greater than the diameter of an atomic nucleus. This was so because the particles carrying the force had substantial mass. For similar reasons, the weak forces, carried by the ponderous W particles (sometimes referred to as "intermediate vector bosons") and Z particles, also have limited range.

Now, all of today's efforts and motivations for seeking this ideal sort of unity come from the hypothesis that at sufficiently high temperatures and energies, all forces are equivalent to one another. Indeed, physicists think that this state of affairs existed at the moment of the big bang. As the universe expanded and cooled, the forces gradually split off from one another. This came to be called "spontaneous symmetry breaking," an idea first introduced in 1974 by Sheldon Glashow and Howard Georgi. (By 1970, Glashow, Luciano Maiani, and John Iliopoulos would already be thinking about fitting quarks, perhaps the ultimate and most basic subatomic particle, into a unified "electroweak" theory of the weak forces and electromagnetism.)

Scientists had known for some time that the photon was the bearer of electromagnetic forces. But Nobel laureate Sheldon Glashow of Harvard asserted confidently, in 1960, that nature used three *more* particles which, he surmised, would carry both the electromagnetic and the weak forces. They would carry the weak, or as he called them, the "electroweak," forces. The problem of the limited range of these $W-$, $W+$, and Z^0 particles, as Glashow dubbed them, still remained, however. Glashow at first tried to accommodate this by suggesting that they had extremely large masses. However, when theoretical scientists introduced these large masses into the mathematics, other absurd results emerged. Certain supposedly "weak" interactions, for example, now turned out to have infinite strength! Glashow was not dispirited; since Richard Feynman had corrected a similar problem with infinities in quantum electrodynamics with a mathematical procedure called "renormalization," Glashow naturally wondered if it might work the same way in this case.

THE ELECTROWEAK FORCES

Although this renormalization procedure too was unsuccessful, physicist Steven Weinberg of the University of Texas at Austin and Abdus Salam of the Imperial College of Science and Technology in London solved the problem of the mass of the particles less than a decade later. In the process, they worked out a theory of a unified electroweak force. Interestingly, this was one of the few instances in the history of science where scientists successfully applied some ideas from the relatively quiet field of solid-state physics to particle physics; the scientific world had long been using spontaneous symmetry breaking in solid-state physics.

Weinberg's particular contribution to the search for the carriers of the "weak" force and, of course, to the unification of the weak force and the electromagnetic force was the "Z^0" particle, which he called a vector boson. Bosons were not new, but this one was uncharged. Weinberg labeled it a "Z" particle. He then predicted that it would appear in the form of the "weak neutral current" at a temperature of 10^{15} degrees. (As opposed to the known "charged" currents, where the exchange of charged, rather than uncharged bosons produces a force.)

Glashow and Weinberg then devised a "test" for determining whether the electromagnetic forces and the weak forces were really merely different manifestations of one and the same force. As it turns out, investigators working at CERN as early as 1973 had seen photographs of neutrinos entering a bubble chamber showing nearly 200 examples of the current predicted by Glashow et al. Still, total confirmation did not come until the discovery of the "weak neutral current" at CERN in 1983, when physicist Carlo Rubbia and his team found the W particle and, very soon afterwards, the Z particle. For this work the Nobel Committee awarded Rubbia the Nobel Prize in 1984. (In an interesting aside, Chinese-American physicist C. N. Yang and Israeli-American physicist Robert Mills, working at the Brookhaven National Laboratory in 1954, tried to apply similar ideas to the strong forces, though this had only limited success.) With the "electroweak" theory proven beyond a reasonable doubt, physics had taken a major step in the direction of Einstein's dream of a unified field theory.

For this work in unifying the electromagnetic and weak forces, Glashow, Salam, and Weinberg shared the 1979 Nobel Prize in physics, "for their contributions to the theory of the unified weak and electromagnetic interaction between elementary particles, including, inter alia, the prediction of the weak neutral current" (a rare instance where the committee awarded the prize before anyone had confirmed the theory—an indication of the confidence the scientific world had in Glashow and Weinberg's work).

As Glashow recalled in the Nobel lecture, "In 1956, when I began doing theoretical physics, the study of the elementary particles was like a patchwork quilt. Electrodynamics, weak interactions, and strong interactions were clearly separate disciplines, separately taught and separately studied. There was no coherent theory that described them all. . . . The patchwork quilt has become a tapestry."

Glashow, not resting on any laurels as so many Nobel Prize winners do, also made contributions to our understanding of the strong forces within the nucleus. Glashow continues to do work on the strong forces today at Harvard, trying to push unification in physics even further via efforts to unite the "electroweak" force with the "strong" force. Weinberg is currently at the University of Texas at Austin and in addition to particle physics, he works in astrophysics and astronomy. Like Glashow, he received his bachelor's degree from Cornell in 1954. He earned his doctorate at Princeton in 1957 for his work on renormalization theory, which was so critical in the development of quantum electrodynamics and particle physics.

In 1983 Rubbia again triumphed, announcing that he had good evidence for still another quark, the heaviest or "top" quark. Then, working at CERN, he found evidence for "monojets," still another kind of particle predicted by supersymmetry theories, though scientists have not yet fully confirmed either of these latter two discoveries.

Indeed, the early 1980s were among the most fruitful years in modern American science. In 1980 at the University of Wisconsin, Joseph Cassinelli discovered the most massive star ever seen—R136A. With over 2500 times the mass of the sun, it was over 100 times brighter. In 1983, William Fowler and Subrahmanyan Chandrasekhar won the Nobel Prize for their seminal work on the collapse of stars. In space technology, the world saw the launching of the first female astronaut into space, Sally K. Ride.

GUTS

Weinberg and Glashow's work had been a step of great consequence toward a GUT or "grand unified theory." But could the unification be extended to include the strong forces? One problem is that science has not yet tested the part of the GUT which includes the "strong" forces, since it requires energies beyond what we are capable of producing on Earth.

If anyone could find a successful GUT, it would constitute a huge stride toward an ultimate explanation of the cosmos. We know, for example, that at the moment of the big bang, there was an infinitesimal excess of matter over antimatter (which is why we live in a universe of matter rather than antimatter). A successful GUT should explain this phenomenon.

GUTs also predict entities known as "magnetic monopoles." Normally understood, a magnet has two poles, north and south. But "monopoles" are essentially particles that are really magnets—but with only a single pole, and this pole would otherwise act as the corresponding pole of an ordinary magnet. If they exist, they are probably extremely massive, even if they are not abundant in the universe. But do they exist? In 1982, physicist Blas Cabrera of Stanford University claimed to have found one, but, so far, no one has substantiated this claim, since physicists have found it immensely hard to repeat the result.

At about the same time Glashow was developing his work on the weak forces, Murray Gell-Mann and George Zweig were working on reducing the number of elementary building blocks of matter. According to what theoretical physicists now call the "standard model," their suggestion was that all hadrons could be thought of as woven out of three distinct and more basic particles called "quarks," as Gell-Mann labeled them. Quarks would then interact via the strong forces. Soon, however, Glashow, purely for theoretical reasons, suggested adding a fourth quark, which he labeled "charm" (He chose this label for no special reason other than a natural inclination among physicists to sound "witty.") Wishing to further validate his views, Glashow again, in 1970, offered considerably more evidence that the charmed quark was real—not just a theoretical fabrication. There were in fact good theoretical reasons for proposing the existence of charmed quarks, since if they existed, an assortment of complications resulting from merely a three-quark theory would sort themselves out. And in 1974, scientists did in fact discover particles carrying charmed quarks, confirming Glashow's prediction.

QUANTUM CHROMODYNAMICS (QCD)

The equations that describe the action of these forces were soon woven into an elegant and bewitching theory known as quantum *chromodynamics* ("chromo" referring to the "colors" of different types of quarks, though the term "color" is not to be taken literally). As the name suggests, the inspiration for quantum chromodynamics was quantum *electrodynamics*. Since the latter had worked so well in describing the electromagnetic forces, it was logical to pattern a theory of the strong forces on it. By analogy with QED then, QCD stipulates that the hadrons are made up of quarks that have a *color*; investigators then suggested that still another particle, called the "gluon," carried the force between quarks. These have the further quality that they bind together the "colors" of two or more quarks, just as pi mesons "glue" together the protons and the neutrons in the nucleus.

And, in 1974, researchers actually found particles with quarks in them. (No one has observed a quark existing separately, however, and certain quark investigators say that this may be impossible.) One of the principal players at this stage in the history of

QCD was MIT physicist Samuel Chao Chung Ting. It was Ting (along with physicist Burton Richter) who discovered the J particle in 1974, thereby adding still further validation to the idea of "charm" inasmuch as the J particle appeared to possess all of the properties predicted by the charm theory. For this he was awarded the 1976 Nobel Prize.

While the above efforts constitute immensely important steps toward a "grand unified field theory" or GUT, difficult problems still remain. For one thing, there is more than one species of a GUT and it is not clear which one is correct. Second, the mathematical technique of "renormalization," despite the wonders it performed in dealing with the recalcitrant problems encountered in QED, remains somewhat suspect. Though it eliminates the infinities in many cases (again as in QED), it does so by a little mathematical treachery, such as dividing one infinity by another—something teachers correctly instruct every grammar schoolchild never to do.

A FIFTH FUNDAMENTAL FORCE IN NATURE?

There is also the distant possibility that someone may find a fifth, yet undiscovered fundamental force. In fact, this almost happened in 1986, when Purdue physicist Ephraim Fischbach made the daring and brilliant suggestion that there might be a fifth force that would act against gravity. If true, that would untangle the experimental discrepancies. Soon, physicists were on a massive hunt to prove or disprove this idea. While a few scientists did get some results vaguely supporting Fischbach, most of the work has produced nothing of significance. Even so, there has apparently been just about enough positive evidence to warrant further exploration. Where it will end is anyone's guess. At the moment, scientists are having enough trouble grappling with uniting four fundamental forces.

CHAPTER 20

Beyond the GUT

Once someone works out a really satisfactory GUT, the next step will be to incorporate *gravity* into the sublime unification scheme. So far, theoreticians have found it onerous in the extreme to unite gravity with the weak, strong, and electromagnetic forces as a genuine "theory of everything" requires. If this could be done, the ultimate crusade of science and philosophy would be over. We would have the "theory of everything" so long sought after by philosophers since the days of the Chinese sage Lao-tzu.

SUPERSYMMETRY

A weighty recent step in the quest is the emergence of an idea called "supersymmetry" or SUSY, first announced in the 1970s. Supersymmetry has a theoretical goal similar to the GUTs and, in fact, is integrally related to the GUT. In supersymmetry, physicists try to wed the seemingly disparate bosons (any particle that carries a force, such as a pi meson, photon, or gluon) and fermions (any particle of matter such as protons, electrons, quarks) into one happy "family." For every fermion there would be a boson, and vice versa.

One other obvious problem in any sort of "unification" scheme has to do not so much with the four forces directly, but with the fact that the *particles* that constitute the physical world are of wildly different types. For one thing, "bosons" and "fermions" have markedly different structures and properties. Then too, their "spin" characteristics differ in bewildering ways. (A larger number of particles spin just like tops. Samuel Goudsmit, for example, discovered the "spinning electron" in an earlier era.) Would it not take some shady trickery to claim that they were the "same" in any sense?

Undeterred, physicists began expanding the idea of supersymmetry even more to include gravity. This resulted in a form of supersymmetry, first suggested in 1976, that included gravity and came to be called "supergravity." Although such new SUSY theories were not vexed by nuisances like the infinities mentioned above, there seemed to be several equally plausible versions of such "supertheories." Which was correct, if any?

Despite all such imaginative schemes, gravity remained intransigent. It could not easily be blended into these new superschemes. As Professor Paul Joss of the physics department of MIT told me, "The fundamental reason is that in order to do so, one has to quantize the space-time field itself." (Put another way, the gravitational field of an object complicates the effort; in addition to attracting other things, the field also attracts *itself*.) Physicists had to show that gravity, like the light emitted when electrons drop from one level to another, is not continuous but exists in discrete packages—hence the idea of the "graviton," by analogy with the photon. The reason one has to quantize gravity is simply that the modern quantum theory of forces *necessarily* involves the exchange of particles or "quanta." Thus, physicists have to talk in terms of gravitational particles or "gravitons." But then one faces the colossal problem of having to try to "quantize" the very medium in which one is working. The great complication is that in addition to the forces gravitons would exert on other particles, physics would have to take into account the forces gravitons exert on *each other*. As difficult as the interaction between, say, hadrons is, at least they "ignore" each other. Gravitons do not.

STRING THEORY

Then, in 1984, a new idea, or rather a dramatically revived old idea, left physicists aghast. Now, along with superparticles, supersymmetry, and supergravity, we have "superstrings." The virtue of this concept was that the problem of incorporating gravity into a unified theory appeared to be solved almost automatically in a "string" description of the world; no artificial efforts to graft it on from the outside were necessary. The main idea in string theories is the jettisoning of the usual technique of conceiving of fundamental particles as mathematical points. Instead, scientists think of them as mathematical "strings"—objects with a finite size but extension in one dimension only. Although the idea had been around since the late 1960s, physicists, for various reasons, had given it only passing attention.

In 1984, however, everything suddenly changed. Michael Green of the University of London and John Schwarz of Caltech proved that the string theory could be wedded with the idea of supersymmetry to provide everything the ultimate "theory of everything" required. Also, it appeared to be free of the annoying mathematical problems discussed above. Nor did it hurt that at this time also, the initially attractive idea of "supergravity" was beginning to fade; it appeared to be ultimately untestable. (For one thing, in supergravity, unlike string theory, particles are still thought of as mathematical points and mathematical points do not exist in nature.)

String theory also had the advantage of being easily visualizable. By analogy with an ordinary piece of string, it is not hard to see that an almost endless variety of motions are possible. A string can rotate, curl up, vibrate, flex, and so forth. The most elementary motion is rotation, where the ends of the string rotate at the speed of light. Such strings, corresponding to what were ordinarily considered "particles," would have the properties of various particles (charge if electromagnetism is involved, "color" if quarks are involved, and so forth). Furthermore, the strings can then join at their ends in "loops," split apart into two new strings, and so forth, in the manner of particles in the older theories. (In fact, as it turns out, string theories without at least a few "loops" don't seem to work.)

Also, string theory could explain why no one has ever found quarks (though they have found material containing them). It may be that quarks are the ends of "strings" which, in turn, are particles. If so, that would explain why quarks are never seen apart from particles, since the "end" of a one-dimensional string will have zero dimensions; that is, it will be a mathematical point and could have no physical reality apart from a particle. Certainly this was one of the most frenetic periods in 20th century science. Many physicists all over the world almost literally dropped everything they were doing to plunge into string theory.

STRINGS AND SPACES

Next, physicists incorporated superstrings into a mosaic geometry of multidimensional spaces. The relevance of "dimensions" to string theory comes from theoretical work done in the 1920s by physicist Theodor Kaluza, who suggested then that spatial dimensions could, under certain circumstances, be thought of as "forces." (And of course "forces" are precisely what physicists are trying to unify.) In this way, they were able to fashion a workable superstring theory with spaces of ten dimensions, rather than eleven as physicists had suggested earlier. The idea of "rolled-up" dimensions is so abstract that it's almost impossible to give it any intuitive plausibility. However, think of placing garbage inside a trash compactor. As the compactor goes to work, the dimensions of the garbage decrease. This is more or less what physicists are talking about when they talk about "compacting" dimensions. The drawback of this approach was that while it simplified and made the mathematics more plausible, it just didn't seem to have much to do with the real world.

COMPACTIFICATION

Of course, physicists tried to explain away the fact that no one ever sees these "extra" dimensions. To do so, they reached into history once again. In the 1920s, Kaluza had discussed a concept called "compactification." In this queer notion, Kaluza claimed that more than the usual four dimensions of space might exist and that they could, in effect, act as forces, as noted above. The problem was that his proof postulated five dimensions rather than the four we normally experience. But since the fifth was invisible, the idea did not grab hold.

Later, however, in 1926, Swedish physicist Oskar Klein showed that such new dimensions could be hidden by being rolled up into a circle or "compacted." (Kaluza and Klein suggested, for example, that the electromagnetic force might be a manifestation of gravity from a "hidden" dimension "leaking" into our own. Klein arrived at this idea in the course of thinking about Kaluza's admission that his fifth dimension is not visible. Klein tried to explain the "hidden" fifth dimension by saying we can't see it because it's all rolled up into an almost infinitesimally small circle.) But unlike a sheet of paper, these extra dimensions could be rolled up to about a size of 10^{-33} centimeters. Again, verification was the problem; things of this size are much too small to be probed with the energies we can generate on Earth. (We might need a particle accelerator the size of the pinwheel galaxy to succeed in this.) It also turned out that several varieties of string theory were imaginable depending on how many dimensions theoretical scientists chose. The idea was so peculiar that virtually no one took it seriously and it soon fell into oblivion.

Then in the late 1970s, Bernard Julia and Eugene Cremmer said physicists had buried the idea prematurely. If one started with the assumption that there were seven hidden dimensions "rolled up," then, for extremely complicated mathematical reasons, the theory seemed to have a little more explanatory power. Japanese-American physicist Nambu, in 1970, offered one of the more electrifying suggestions. Nambu's original theory had twenty-six dimensions, though physicists now believe it is not necessary to postulate more than ten or so.

But do these theories also contain troublesome infinities? Infinities often arise because physicists find it necessary to conceive of a particle such as an electron as having zero size. But the string theories sidestep this difficulty by regarding particles as having a small, but nonzero size—a length of about 10^{-36} meters. With that way of looking at things, chances of infinities were greatly diminished and so far, none have appeared. Nevertheless, it is still thinkable that they may. The problem is that the mathematics of string theory is so inscrutable that physicists must make approximations, using another mathematical technique called "perturbation theory." At the levels of estimation so far examined, the frightening infinities have not shown up. As for the future, no one can tell.

And even if physicists escape that fate, other perils loom. For one, as with supergravity, there are a number of feasible theories and even a vast number of possible varieties of the ones that scientists have already proposed (although just two of these escape the infinities that haunted QED and the GUTs, thereby narrowing the possibilities considerably). Also, the idea that a number of the extra dimensions can be "rolled up" is a fantasy so bizarre that features like these have led even adventuresome physicists like Richard Feynman to say that string theories are "nonsense." Feynman may be right, since other researchers such as physicist Edward Witten of Princeton have offered some reason to believe that postulating several "hidden" dimensions may not be inevitable—our usual four just might suffice.

But even such "compactification" of the extra dimensions drives physicists into apparently unanswerable, almost theological speculation. Why are all of those extra dimensions compressed like this? Did this have something to do with the creation of the universe? Have physicists been watching too many sci-fi movies?

On the other hand, the idea of "rolled-up dimensions" hidden from our view would seem to allow the possibility of extra matter hidden from our view—an exceedingly useful idea for physicists working on the question of whether the universe will eventually tumble down on itself or not. According to cosmologists, there are three possibilities: either the universe will expand forever, eventually collapse, or it will "flatten." Relevant here is the "inflationary universe" idea suggested by Alan Guth in 1979. According to Guth, the universe had more energy and expanded far faster after the big bang than we have always supposed. If so, then there is far more matter around today than physicists have supposed. The reason is that according to well-known principles in quantum mechanics, energy can be converted into matter. Thus, there is less "missing" matter today that would be needed to close or at least flatten the universe. Still, far more mass is needed than we can actually find.

Even so, in 1980, the prospects for finding the missing mass skyrocketed when a number of research teams, working independently, offered good reasons for believing that Fermi's "little neutral ones" or neutrinos might have a minuscule mass. If so, that could well end the hunt.

Perhaps most important of all, the real test of any viable scientific theory is whether it produces confirmable predictions. So far at least, the number of these has been extremely small and even some of the experimental tests that physicists have carried out contradict the facts. For example, superstring theory predicts the existence of particles with the mass of a particle of lint—unbelievably and absurdly large for a subatomic particle. In fact, there is no experimental evidence whatsoever for the existence of particles that large.

This is not to say that all hope for string theory is lost. New work may bolster the ideas at some point in the future, though many remain pessimistic. Some, however, remain guardedly hopeful. As Nobel Prize winner Steven Weinberg says, "Whether string theory will have been a good idea depends on what comes out of it . . . it's certainly going to be a lot of fun in the next few years to work that out."

Endnotes

CHAPTER 1

1. Robert Post, "Science, Public Policy, and Popular Precepts: Alexander Dallas Bache and Alfred Beach as Symbolic Adversaries," in Nathan Reingold, editor, *The Sciences in the American Context: New Perspectives* (Smithsonian Institution Press, Washington, D. C., 1979), p. 82.
2. Stanley Guralnick, "The American Scientist in Higher Education, 1820–1910," in Nathan Reingold, editor, *The Sciences in the American Context: New Perspectives* (Smithsonian Institution Press, Washington, D.C., 1979), p. 188.
3. Deborah Jean Warner, "Astronomy in Antebellum America," in Nathan Reingold, editor, *The Sciences in the American Context: New Perspectives* (Smithsonian Institution Press, Washington, D.C., 1979), p. 62.
4. Carlene Stephens, "Partners in Time: William Bond & Son of Boston and the Harvard College Observatory," delivered at exhibition commemorating 150th anniversary of Harvard College Observatory, organized by Department of the History of Science and Technology, National Museum of American History, and the Smithsonian Institution, held at the Collection of Historical Scientific Instruments, Harvard University, January 10 through June 9, 1989.

5. Mitchell Wilson, *American Science and Invention* (Simon & Schuster, New York, 1954), p. 274; see also Carlene Stephens, op. cit.

6. Ibid.

7. Annie J. Cannon, "Classifying the Stars," from Harlow Shapley and Cecilia H. Payne, editors, *The Universe of Stars*, as reprinted in Timothy Ferris, editor, *The World Treasury of Astronomy and Physics* (Little, Brown, Boston, 1991), pp. 272–273.

8. K. M. Olesko, "Michelson and the Reform of Physics Instruction at the Naval Academy in the 1870's," in Stanley Goldberg and Roger Stuewer, editors, *The Michelson Era in American Science: 1870–1930* (American Institute of Physics, New York, 1988).

9. George Biddell Airy, "Account of some circumstances historically connected with the discovery of the planet exterior to Uranus," *Monthly Notices of the Royal Astronomical Society*, 1846, 7:124. Interestingly, according to W. H. Pickering in his article "A Search for a Planet Beyond Neptune" (*Annals of the Astronomical Observatory of Harvard College*, Vol. LXI, Part II, p. 133), Bouvard appears to have hindered the search because of some erroneous calculations.

10. Benjamin Peirce, "On the Law of Vegetable Growth and the Periods of the Planets," *Proceedings of the American Academy of Arts and Sciences*, 1852, 2:241.

11. George Forbes, "On Comets," *Proceedings of the Royal Society of Edinburgh*, 1880, 10:427; W. H. Pickering, in his article "The Transneptunian Planet" (*Annals of the Harvard College Observatory*, Vol. 82, No. 3), asserts that Forbes "located it [planet beyond Neptune] by means of the orbits of certain comets."

12. Todd to Forbes, personal letter of June 15, 1880, in "Additional Note on an ultra-Neptunian Planet," *Proceedings of the Royal Society of Edinburgh*, 1882, 11:91–92.

13. J. C. Kapteyn to G. E. Hale, personal letter of February 7, 1919, from Harvard University Archives, Lawrence Lowell Papers, UAI 5.160 (1919–22); see also Owen Gingerich, "How Shapley Came to Harvard, Or, Snatching the Prize from the Jaws of Debate," JHA, 1988, xix.

14. Hale to Lowell, personal letter of March 12, 1920, also from Harvard Archives/Lowell Papers.

15. G. R. Agassiz to Lowell, personal letter of March 19, 1920, also from Harvard Archives/Lowell Papers.

16. Hale to Lowell, personal letter of March 29, 1920, also from Harvard Archives/Lowell Papers.

CHAPTER 2

1. Rowland to his mother, May 1868, Rowland and Gilman Collections at Johns Hopkins University.
2. Rowland to his mother, February 23, 1875, Rowland and Gilman Collections at Johns Hopkins University.
3. I am informed by David Miller (private correspondence) that some are at the US Museum of History and Technology; also, David Miller compares Rowland's collection with the Harvard collection in his dissertation, "Henry Augustus Rowland and his Electromagnetic Researches," for Ph.D. in General Science, thesis submitted to Oregon State University, June 1970, pp. 292 and 293.
4. From Miller's dissertation, op cit., pp. 289–292.
5. J. Ames, "The Work of the Physical Laboratory," *Johns Hopkins University Alumni Magazine*, 1917, 5:158.
6. C. J. Evans and D. J. Warner, "Precision Engineering and Experimental Physics: William A. Rogers, the First Academic Mechanician in the U.S.," in Stanley Goldberg and Roger Stuewer, editors, *The Michelson Era in American Science: 1870–1930* (American Institute of Physics, New York, 1988).
7. David Miller, "Rowland and the Nature of Electric Currents," *Isis*, 1972, 63:5–27.
8. Rowland to Helmholtz, correspondence, November 13, 1875, in David Miller, "Rowland and the Nature of Electric Currents," *Isis*, 1972, 63:5–27.
9. David Miller, "Rowland and the Nature of Electric Currents," *Isis*, 1972, 63:5–27; see also David Miller, "Rowland's Physics," *Physics Today*, July 1976.
10. Anthony to Board of Trustees, June 19, 1887, Cornell University Archives.
11. Rowland to Fitzgerald, May 12, 1876, Royal Dublin Society Library.

CHAPTER 3

1. Lynde Phelps Wheeler, *Josiah Willard Gibbs, The History of a Great Mind* (Yale University Press, New Haven, Conn., 1951), p. 21.
2. "One of the Prophets," Margaret Whitney, read to Saturday Morning Club of New Haven.

3. George E. Uhlenbeck, "Fifty Years of Spin," in *Readings From Physics Today*, No. 2 (American Institute of Physics, New York, 1985).
4. Hertz to Gibbs, March 3, 1889.
5. Pattison Muir to Gibbs, February 14, 1880.
6. Rowland to Gibbs, March 3, 1879.
7. Mitchell Wilson, *American Science and Invention* (Simon & Schuster, New York, 1954), p. 298.
8. Lynde Phelps Wheeler, *Josiah Willard Gibbs, the History of a Great Mind* (Yale University Press, New Haven, Conn., 1951), p. 99 (Wheeler's translation).
9. Daniel Kevles, *The Physicists* (Alfred A. Knopf, New York, 1978), p. 34.
10. Lynde Phelps Wheeler, *Josiah Willard Gibbs, the History of a Great Mind* (Yale University Press, New Haven, Conn., 1951), p. 91.
11. Gibbs to Schlegel, August 1, 1888.

CHAPTER 4

1. A. A. Michelson, "Some of the Objects and Methods of Physical Science," *Quarterly Calendar*, August 1884.
2. Elizabeth Crawford, "Scientific Elite Revisited: American Candidates for the Nobel Prizes in Physics and Chemistry 1901–1938," in *The Michelson Era in American Science: 1870–1930* (American Institute of Physics, New York, 1988), p. 260.
3. From Dorothy Michelson Livingston, *The Master of Light: A Biography of Albert A. Michelson* (Scribner's, New York, 1973).
4. K. M. Olesko, "Michelson and the Reform of Physics Instruction at the Naval Academy in the 1870s," in *The Michelson Era in American Science: 1870–1930* (American Institute of Physics, New York, 1988), p. 43.
5. This is thoroughly discussed by William Koelsch in his article "The Michelson Era at Clark, 1889–1892," in *The Michelson Era in American Science: 1870–1930* (American Institute of Physics, New York, 1988), pp. 133–151.
6. Mitchell Wilson, *American Science and Invention* (Simon & Schuster, New York, 1954), p. 311.
7. Robert Millikan, *The Autobiography of Robert Millikan* (Prentice–Hall, Englewood Cliffs, N.J., 1950), p. 87; compare this version with "Albert

Abraham Michelson," *Biographical Memoirs of the National Academy of Sciences*, 1938, p. 126.

8. Lynde Phelps Wheeler, *Josiah Willard Gibbs, the History of a Great Mind* (Yale University Press, New Haven, Conn., 1951), pp. 140–141.

9. Mitchell Wilson, *American Science and Invention* (Simon & Schuster, New York, 1954), p. 314.

10. A. A. Michelson, "The Relative Motion of the Earth and the Luminiferous Ether," *American Journal of Science*, 1881, XXII.

11. Daniel J. Kevles, *The Physicists* (Alfred A. Knopf, New York, 1978), p. 29.

12. H. A. Lorentz, "The Relative Motion of the Earth and the Ether," *V. K. Akad. Wet. Amsterdam*, 1892, 1:74; in *H. A. Lorentz, Collected Papers*, Vol. 4, Zeeman and Foker, editors, pp. 219–223; see also J. Z. Buchwald, "The Michelson Experiment in the Light of Electromagnetic Theory Before 1900," in *The Michelson Era in American Science: 1870–1930* (American Institute of Physics, New York, 1988).

13. Barbara Haubold *et al.*, "Michelson's First Ether Drift Experiment in Berlin and Potsdam," in *The Michelson Era in American Science: 1870–1930* (American Institute of Physics, New York, 1988).

14. J. Z. Buchwald, "The Michelson Experiment in the Light of Electromagnetic Theory Before 1900," in *The Michelson Era in American Science: 1870–1930* (American Institute of Physics, New York, 1988).

15. Lloyd S. Swenson, "Michelson–Morley, Einstein and Interferometry," in *The Michelson Era in American Science: 1870–1930* (American Institute of Physics, New York, 1988), p. 238.

16. A. Einstein, translated by Y. A. Ono, "How I Created the Theory of Relativity," *Physics Today*, August 1982.

17. D. H. Stapleton, "The Context of Science: The Community of Industry and Higher Education in Cleveland in the 1880s," in *The Michelson Era in American Science: 1870–1930* (American Institute of Physics, New York, 1988), p. 21.

CHAPTER 5

1. R. A. Millikan, "Juvenalia," Robert A. Millikan Papers, California Institute of Technology, Pasadena, California.

2. Robert Millikan, *The Autobiography of Robert Millikan* (Prentice–Hall, Englewood Cliffs, N.J., 1950), p. 37.

3. 1964 interview with Linus Pauling, by John Heilbron.
4. Robert Millikan, *The Autobiography of Robert Millikan* (Prentice–Hall, Englewood Cliffs, N.J., 1950), p. 62.
5. Mitchell Wilson, *American Science and Invention* (Simon & Schuster, New York, 1954), p. 334.
6. John L. Michel, "The Chicago Connection: Michelson and Millikan, 1894–1921," in *The Michelson Era in American Science: 1870–1930* (American Institute of Physics, New York, 1988), p. 167.
7. James Trefil, *From Atoms to Quarks* (Scribner's, New York, 1980), p. 22.
8. Robert Millikan, *The Autobiography of Robert Millikan* (Prentice–Hall, Englewood Cliffs, N.J., 1950), p. 100.
9. J. L. Michel, "The Chicago Connection: Michelson and Millikan, 1894–1921," in *The Michelson Era in American Science: 1870–1930* (American Institute of Physics, New York, 1988).
10. R. A. Millikan, "Quantum Relations in Photo-electric Phenomena," *Proceedings of the National Academy of Sciences*, 1916, 2:78.
11. Robert H. Kargon, *The Rise of Robert Millikan* (Cornell University Press, Ithaca, N.Y., 1982), p. 72, where Kargon discusses Stuewer's comments on this claim.
12. Daniel J. Kevles, *The Physicists* (Alfred A. Knopf, New York, 1978), p. 113.
13. Joseph Boyce to John Cockcroft, personal letter of 8 January, in the Sir John Cockcroft Papers, Churchill College Library, Cambridge, England.
14. Charles Weiner, "1932—Moving into New Physics," *Physics Today*, May 1972.
15. R. A. Millikan, "Science and Life," 1924, p. 68, Robert A. Millikan Papers, California Institute of Technology, Pasadena.

CHAPTER 6

1. See, for example, David H. DeVorkin, "Steps toward the Hertzsprung–Russell Diagram," *Physics Today*, March 1978.
2. J. C. Kapteyn to George Ellery Hale, personal letter, February 7, 1919, from the A. Lawrence Lowell Papers, UAI 5.160 (1919–22) #14 Observatory.

3. Hale to Lowell, March 29, 1920, from the A. Lawrence Lowell Papers, UAI 5.160 (1919–22) #14 Observatory.
4. Owen Gingerich, "How Shapley Came to Harvard, Or, Snatching the Prize from the Jaws of Debate," JHA, 1988, xix.
5. W. H. Pickering credits a suggestion by Herschel as helping Pickering find Pluto on the very first page of Pickering's article "The Transneptunian Planet" (*Annals of the Harvard College Observatory*, Vol. 82, No. 3). For a good general discussion of Herschel's contributions to American astronomy, see Deborah Jean Warner, "Astronomy in Antebellum America," in Nathan Reingold, editor, *The Sciences in the American Context: New Perspectives* (Smithsonian Institution Press, Washington, D.C., 1979).
6. See Mitchell Wilson, *American Science and Invention* (Simon & Schuster, 1954), p. 275, for a discussion of Clark's role in American astronomy; for an excellent discussion of astronomical technology in this era, see Carlene Stephens, "Partners in Time: William Bond & Son of Boston and the Harvard College Observatory," essay accompanying exhibition held at Harvard University (from January 10 through June 9, 1989).

CHAPTER 7

1. From the Introduction to Langmuir's book *Phenomena, Atoms and Molecules* (Philosophical Library, 1950), p. viii.
2. Ibid.
3. For an excellent discussion of Langmuir's contributions and Whitney's relationship to Langmuir, see Willis R. Whitney, "Irving Langmuir, Scientist," *Current History*, March 1933.
4. Jeffrey Sturchio *et al.*, *Chemistry in America, 1876–1976* (Reidel, Dordrecht, 1983), p. 111.
5. Marjorie Johnston, editor, *The Cosmos of Arthur Holly Compton* (Alfred A. Knopf, New York, 1967), p. 240.
6. Daniel J. Kevles, *The Physicists* (Alfred A. Knopf, New York, 1978), p. 100; also, physical chemist Verner Schomaker discussed his role and early work in electron diffraction at Caltech with me at his office at the University of Washington in 1984.
7. Willis R. Whitney, "Irving Langmuir, Scientist," *Current History*, March 1933.

8. For a brief discussion of this controversy, see my *Linus Pauling: A Man and His Science* (Paragon House, New York, 1989), Chapter 6.
9. Willis R. Whitney, "Irving Langmuir, Scientist," *Current History*, March 1933.
10. As quoted in Willis R. Whitney, "Irving Langmuir, Scientist," *Current History*, March 1933 (Whitney does not give a primary source).
11. Jeffrey Sturchio *et al.*, *Chemistry in America, 1876–1976* (Reidel, Dordrecht, 1983), p. 457; in 1929, toward the end of his life, Langmuir was honored by being appointed to the prestigious post of president of the American Chemical Society.
12. For a brief discussion of Langmuir and quantum theory, see Katherine Russell Sopka, *Quantum Physics in America* (American Institute of Physics, New York, 1985), p. 105.

CHAPTER 8

1. Katherine R. Sopka, *Quantum Physics in America* (American Institute of Physics, New York, 1988), Vol. 10, p. 277.
2. Pauling to W. A. Noyes, January 26, 1938, Archives of the University of Illinois.
3. Interview with Linus Pauling, by John Heilbron, Pasadena, March 27, 1964.
4. Interview with Linus Pauling, by John Heilbron, Pasadena, March 27, 1964.
5. Linus Pauling, "Fifty Years of Progress in Structural Chemistry and Molecular Biology," *Daedalus (Boston)*, May 1970.
6. Based on information kindly provided to me by Linus Pauling, in a letter from Linus Pauling to author, September 30, 1989.

CHAPTER 9

1. *The Physical Review*, Vol. 21, 1923.
2. According to Compton in a letter of August 11, 1949, to Samuel Glasstone, the author of *Sourcebook on Atomic Energy*, G N. Lewis of the University of California first used the word "photon" to describe a

quantum of energy emitted by one atom and absorbed by another; see Arthur Holly Compton, *The Cosmos of Arthur Holly Compton*, Marjorie Johnston, editor (Alfred A. Knopf, New York, 1967), p. 43.

3. Katherine Russell Sopka, *Quantum Physics in America* (American Institute of Physics, New York, 1988), Vol. 10, p. 91.

4. Arthur Holly Compton, *The Cosmos of Arthur Holly Compton*, Marjorie Johnston, editor (Alfred A. Knopf, New York, 1967), p. 107.

5. W. H. Bragg, letter to *Nature*, May 27, 1915; see also A. H. Compton, "A Quantum Theory of the Scattering of X-rays by Light Elements," *The Physical Review*, May 1923, Vol. 21, No. 5.

6. Arthur Holly Compton, *The Cosmos of Arthur Holly Compton*, Marjorie Johnston, editor (Alfred A. Knopf, New York, 1967), p. 144.

7. See E. U. Condon, "60 Years of Quantum Mechanics," *Physics Today*, October 1962.

8. John Slater to Linus Pauling, personal letter, December 16, 1953.

9. See Daniel J. Kevles, *The Physicists* (Alfred A. Knopf, New York, 1978), p. 240.

10. Nathan Reingold, "Physics and Engineering in the United States, 1945–1965, A Study of Pride and Prejudice," in *The Michelson Era in American Science: 1870–1930* (American Institute of Physics, New York, 1988), p. 295.

11. Katherine Russell Sopka, *Quantum Physics in America* (American Institute of Physics, New York, 1988), Vol. 10, p. 147.

12. A. Russo and M. De Maria, "Cosmic Ray Romancing: The Discovery of the Latitude Effect and the Compton–Millikan Controversy," in *Historical Studies in the Physical and Biological Sciences* (University of California Press, Berkeley, 1986), pp. 211–216.

13. Arthur Holly Compton, *The Cosmos of Arthur Holly Compton*, Marjorie Johnston, editor (Alfred A. Knopf, New York, 1967), p. 43.

14. Arthur Holly Compton, *The Cosmos of Arthur Holly Compton*, Marjorie Johnston, editor (Alfred A. Knopf, New York, 1967), p. 86.

15. A. Russo and M. De Maria, "Cosmic Ray Romancing: The Discovery of the Latitude Effect and the Compton–Millikan Controversy," in *Historical Studies in the Physical and Biological Sciences* (University of California Press, Berkeley, 1986), pp. 211–216.

16. J. Robert Oppenheimer, to his brother Frank, personal letter, June 4, 1939.

17. Silvan S. Schweber, "Shelter Island, Pocono, and Oldstone: The

Emergence of American Quantum Electrodynamics after World War II," *Osiris*, Vol. 2, 1986.

18. Laurie M. Brown and Lillian Hoddeson, "The Birth of Elementary Particle Physics," *Physics Today*, April 1982; see also Robert Marshak, "The Multiplicity of Particles, as reprinted in *Scientific American Reader* (Simon & Schuster, New York, 1953).

19. Laurie M. Brown and Lillian Hoddeson, "The Birth of Elementary Particle Physics," *Physics Today*, April 1982.

20. G. Lemaitre, *The Primeval Atom*, Betty Korff and Serge Korff, translators, 1950.

21. Arthur Holly Compton, *The Cosmos of Arthur Holly Compton*, Marjorie Johnston, editor (Alfred A. Knopf, New York, 1967), p. 430.

22. Arthur Holly Compton, *The Cosmos of Arthur Holly Compton*, Marjorie Johnston, editor (Alfred A. Knopf, New York, 1967), p. 439.

23. Willard Libby to Secretary of Commerce Lewis Strauss, personal letter, November 25 (from the files of the AEC).

24. Arthur Holly Compton, *The Cosmos of Arthur Holly Compton*, Marjorie Johnson, editor (Alfred A. Knopf, New York, 1967).

CHAPTER 10

1. Richard K. Gehrenbeck, "Electron Diffraction: Fifty Years Ago," *Physics Today*, January 1978 (no primary source given).

2. See Jeremy Bernstein, *Prophet of Energy: Hans Bethe* (E. P. Dutton, New York, 1981), pp. 18–19.

3. Lester H. Germer, "The Structure of Crystal Surfaces," *Scientific American*, June 1963.

4. Notebook entry of February 5, 1925.

5. C. J. Davisson, "Are Electrons Waves?" *Franklin Institute Journal*, 1928, 105:597–623.

6. Richard K. Gehrenbeck, "Electron Diffraction: Fifty Years Ago," *Physics Today*, January 1978.

7. Lester H. Germer, "The Structure of Crystal Surfaces," *Scientific American*, June 1963.

8. For an excellent discussion of this, see Katherine Russell Sopka, *Quantum Physics in America* (American Institute of Physics, New York, 1985), Vol. 10, p. 87.

CHAPTER 11

1. This famous quote is reproduced in "Physics in the Great Depression," Charles Weiner, *Physics Today*, October 1970, who quotes from the *Herald Tribune* of September 12, 1933.
2. Luis W. Alvarez, "Alfred Lee Loomis—the Last Great Amateur of Science," *Physics Today*, January 1982.
3. E. O. Lawrence to Linus Pauling, October 3, 1947, Archives of the Lawrence Radiation Laboratory.
4. Cablegram from Lawrence to Pauling, November 5, 1954, Archives of the Lawrence Radiation Laboratory.
5. Lawrence discusses the cyclotron in E. O. Lawrence and M. S. Livingston, "Production of High Speed Ions," *The Physical Review*, 1932, 42:20–35; at the end of this paper, Lawrence mentions the possibility of stronger cyclotrons with guarded optimism.
6. Lawrence to Gamow, December 27, 1933, as quoted by Mark L. Oliphant, "The Two Ernests—I," *Physics Today*, September/October 1966.
7. Lawrence to Cockcroft, personal letter of Jan 12, 1924, in Mark L. Oliphant, "The Two Ernests—I," *Physics Today*, September/October 1966.
8. See Lise Meitner and O. R. Frisch, "Disintegration of Uranium by Neutrons: A New Type of Nuclear Reaction," *Nature*, 1939, 143:239f; interestingly, the authors claim that the entire fission process can be described using only classical physics.
9. See E. O. Lawrence, "Amateur of the Sciences," *Fortune*, March 1946.
10. See I. Joliot-Curie and J. Joliot-Cure, "A New Type of Radioactivity," *Comptes Rendus*, 1934, p. 198.
11. See Chamberlin, Segrè, Wiegand, and Ypsilantis, "Antiprotons," *Nature*, 1956, 177, where the authors confidently predict having found the antiproton, saying it is a "virtual certainty that the antineutron exists."
12. Lawrence to Rutherford, March 1936, as quoted in Mark L. Oliphant, "The Two Ernests—II," *Physics Today*, September/October 1966.

CHAPTER 12

1. Interview with Robert Bacher, June 1984, California Institute of Technology.

2. P. A. M. Dirac, Nobel Prize Address, 1933, as reprinted in Timothy Ferris, editor, *The World Treasury of Physics, Astronomy and Mathematics* (Little, Brown, Boston, 1991), p. 83.

3. P. A. M. Dirac, Nobel Prize Address, 1933, as reprinted in Timothy Ferris, editor, *The World Treasury of Physics, Astronomy and Mathematics* (Little, Brown, Boston, 1991), p. 84.

4. See Laurie Brown and Lillian Hoddeson, "The Birth of Elementary-Particle Physics," *Physics Today*, April 1982.

5. Ibid.

6. See Laurie M. Brown, "The Idea of the Neutrino," *Physics Today*, September 1978, for more on theoretical work on the nucleus during this period.

7. See also Daniel J. Kevles, *The Physicists* (Alfred A. Knopf, New York, 1978), p. 231.

8. From Robert Millikan to J. Robert Oppenheimer, personal letter of August 31 (from the Archives of the California Institute of Technology).

9. See Laurie Brown and Lillian Hoddeson, "The Birth of Elementary-Particle Physics," *Physics Today*, April 1982.

10. Interview of Linus Pauling by John Heilbron, March 27, 1964.

11. Daniel J. Kevles, *The Physicists* (Alfred A. Knopf, New York, 1978), p. 233.

12. Alice Kimball Smith and Charles Weiner, "The Young Oppenheimer: Letters and Recollections," *Physics Today*, April 1980.

13. Hans Bethe and Walter Heitler, *Proceedings of the Royal Society of London Series A*, 1934, 146:83.

CHAPTER 13

1. Samuel A. Goudsmit, "It Might as Well be Spin," *Physics Today*, June 1976.

2. H. A. Bethe, "Energy Production in Stars," *The Physical Review*, 1938, 55. As reproduced in Henry A. Boorse and Lloyd Motz, *The World of The Atom*, Vol. II (Basic Books, New York, 1966).

3. See Jeremy Bernstein, *Prophet of Energy: Hans Bethe* (E. P. Dutton, New York, 1981), Chapter 2; see also Henry A. Boorse and Lloyd Motz, *The World of the Atom*, Vol. II (Basic Books, New York, 1966), Chapter 90.

4. H. A. Bethe, "Energy Production in Stars," *The Physical Review*, 1938, 55.

CHAPTER 14

1. See Niels Bohr, "On the Constitution of Atoms and Molecules," *Philosophical Magazine*, Series 6, 1913 for Bohr's early thoughts on atomic structure.
2. Enrico Fermi, "Experimental Production of a Divergent Chain Reaction," *American Journal of Physics*, 1952, p. 20; in this article Fermi discusses the structure, operation, and energy production in an atomic pile.
3. Urey had discussed isotope separation as early as 1932, in the classic paper, Urey, Brickwedde, and Murphy, "A Hydrogen Isotope of Mass 2 and Its Concentration," *The Physical Review*, 1932, 40; here he discusses separation of hydrogen isotopes by diffusion.
4. For more on Oppenheimer's college and graduate years, see Alice Kimball Smith and Charles Weiner, "The Young Oppenheimer: Letters and Recollections," *Physics Today*, April 1980; see also Katherine Russell Sopka, *Quantum Physics in America* (American Institute of Physics, New York, 1988), Vol. 10, p. 169.
5. Carl D. Anderson, "The Positive Electron," *The Physical Review*, 1933, p. 43.
6. Alice Kimball Smith and Charles Weiner, "The Young Oppenheimer: Letters and Recollections," *Physics Today*, April 1980.
7. Marjorie Johnston, editor, *The Cosmos of Arthur Holly Compton* (Alfred A. Knopf, New York, 1967), p. 267.
8. The Bird Dogs, "The Evolution of the Office of Naval Research," *Physics Today*, August 1961.

CHAPTER 15

1. Henry A. Boorse and Lloyd Motz, *The World of the Atom*, Vol. II (Basic Books, New York, 1966), p. 1722.
2. For a discussion of symmetry in molecular biology, see Anthony Serafini, *Linus Pauling: A Man and His Science* (Paragon House, New York, 1989), Chapter 6.

CHAPTER 16

1. For a good discussion of solid-state physics in antiquity, see Gregory H. Vannier, "The Solid State," as reproduced in *The Scientific American Reader* (Simon & Schuster, New York, 1953), pp. 140–152.
2. Pauling interview by John Heilbron, Pasadena, March 27, 1964, from the Niels Bohr Library, American Institute of Physics.
3. Probably the best discussion of Van Vleck's work is P. W. Anderson, "Van Vleck and Magnetism," *Physics Today*, October 1968.
4. J. H. Van Vleck, "Reminiscences of the First Decade of Quantum Mechanics," *International Journal of Quantum Chemistry*, 1971, p. 5.
5. E. U. Condon and P. M. Morse, *Quantum Mechanics* (McGraw–Hill, New York, 1929).
6. Nobel Prize Winners (H. W. Wilson, New York, 1987), p. 150.
7. P. W. Anderson to author, private correspondence, September 21, 1988.
8. To Peter Adams, editor of *The Physical Review*, 1984, personal letter.
9. J. Slater, "Electrical Energy Bands in Metals," *The Physical Review*, 1934, p. 45.
10. J. Slater, "The Theory of Complex Spectra," *The Physical Review*, 1929, p. 34.
11. Slater to an unidentified person, personal letter of April 11, 1931.
12. W. A. Little, "Superconductivity at Room Temperature," *Scientific American*, February 1965.
13. Roland W. Schmitt, "The Discovery of Electron Tunneling into Superconductors," *Physics Today*, December 1961.
14. See John C. Slater, "Energy Bands in Solids," *Physics Today*, April 1968; for a striking example of Slater's strong views on the role of computation in problems in solid-state physics, see also John C. Slater, "Quantum Physics in America Between the Wars," *Physics Today*, January 1968.
15. For a discussion of Shockley's work, see Lillian Hartmann Hoddeson, "The Roots of Solid-State Research at Bell Labs," *Physics Today*, March 1977.

CHAPTER 17

1. See Laurie Brown and Lillian Hoddeson, "The Birth of Elementary-Particle Physics," *Physics Today*, April 1982, for an early discussion of

quantum electrodynamics (QED), particularly with respect to Dirac's work.

2. H. Yukawa, "On the Interaction of Elementary Particles," from Yukawa's *Progress of Theoretical Physics* (Kyoto, Tokyo, 1935). See also H. Yukawa, *Proceedings of the Physics and Mathematics Society of Japan*, 1935, 17:48.

3. See Carl Anderson, *Science*, 1932, 76:238, and Neddermeyer and Anderson, *The Physical Review*, 1937, 57:884.

4. See also Daniel J. Kevles, *The Physicists* (Alfred A. Knopf, New York, 1978), p. 231, for a good discussion of Millikan's cosmic ray studies with Anderson.

5. For a useful discussion of SLAC and particle research, see Steven Weinberg, *The Discovery of Subatomic Particles* (Scientific American Books, New York, 1983).

6. See James S. Trefil, *From Atoms to Quarks* (Scribner's, New York, 1980), p. 166, for a good discussion of Rank's work.

7. First suggested by Pauli at a conference on radioactivity in 1930; see also Laurie M. Brown, "The Idea of the Neutrino," *Physics Today*, September 1978.

8. Laurie M. Brown, "The Idea of the Neutrino," *Physics Today*, September 1978.

9. Niels Bohr was virtually a professional skeptic; for his skepticism on positrons, see Daniel J. Kevles, *The Physicists* (Alfred A. Knopf, New York, 1978), p. 233.

10. See Alice Kimball Smith and Charles Weiner, "The Young Oppenheimer: Letters and Recollections," *Physics Today*, April 1980.

11. For a discussion of Einstein–Bose statistics, again see Yukawa's famous paper, "On the Interaction of Elementary Particles," from Yukawa's *Progress of Theoretical Physics* (Kyoto, Tokyo, 1935).

CHAPTER 18

1. Richard Feynman, "The Relation of Science and Religion," *Engineering and Science*, June 1956.

2. See Martin J. Klein, "Thermodynamics and Quanta in Planck's Work," *Physics Today*, November 1966.

3. MacInnes Diary, May 20, 1947, MacInnes Papers, Box 9, Rockefeller University Archives, 450, M189, New York.

4. P. A. M. Dirac, "The Relativistic Theory of the Electron," *Proceedings of the Royal Society*, London, 1928.
5. Willis Lamb, "Fine Structure of the Hydrogen Atom," *Science*, 1956, p. 123.
6. Feynman discusses the relation of his work to Schwinger's, for example, in his article "Space-Time Approach to Quantum Electrodynamics," *The Physical Review*, 1949, 76:769–774.
7. H. A. Kramers, *Collected Scientific Papers* (Amsterdam, 1956).
8. Though Schwinger offers a relatively technical, although non-mathematical discussion of quantum electrodynamics in his article "Quantum Electrodynamics," from *Selected Papers in Quantum Electrodynamics* (Dover, New York, 1958).
9. Silvan S. Schweber, "Shelter Island, Pocono, and Oldstone: The Emergence of American Quantum Electrodynamics after World War II, *Osiris*, 1986, p. 2.
10. See M. Gell-Mann and E. P. Rosenbaum, "Some Strange Particles," *Scientific American*, 1957, p. 197.

Index